ALLE ZEIT WACH
1842

Beethoven 9

Bach 131

Kenneth G. Denbigh

Three Concepts of Time

Springer-Verlag
Berlin Heidelberg New York 1981

Professor Kenneth G. Denbigh, F.R.S.
Council for Science and Society
3/4 St. Andrew's Hill
GB-London EC4V 5BY

ISBN 3-540-10757-6 Springer-Verlag Berlin Heidelberg New York
ISBN 0-387-10757-6 Springer-Verlag New York Heidelberg Berlin

Library of Congress Cataloging in Publication Data. Denbigh, Kenneth George. Three concepts of time. Includes bibliographical references and index. 1. Time. I. Title. BD638.D43. 115. 81-5684. ISBN 0-387-10757-6 (U.S.). AACR2

Printing and binding: K. Triltsch, Würzburg
2141/3140-543210

Preface

The existence of so many strangely puzzling, even contradictory, aspects of 'time' is due, I think, to the fact that we obtain our ideas about temporal succession from more than one source – from inner experience, on the one side, and from the physical world on the other. 'Time' is thus a composite notion and as soon as we distinguish clearly between the ideas deriving from the different sources it becomes apparent that there is not just one time-concept but several. Perhaps they should be called variants, but in any case they need to be seen as distinct. In this book I shall aim at characterising what I believe to be the three most basic of them. These form a sort of hierarchy of increasing richness, but diminishing symmetry.

Any adequate inquiry into 'time' is necessarily partly scientific and partly philosophical. This creates a difficulty since what may be elementary reading to scientists may not be so to philosophers, and vice versa. For this reason I have sought to present the book at a level which is less 'advanced' than that of a specialist monograph.

Due to my own background there is an inevitable bias towards the scientific aspects of time. Certainly the issues I have taken up are very different from those discussed in several recent books on the subject by philosophers. Nevertheless much emphasis is given to the origins of the time-concept in conscious awareness for I believe that the relationship between 'introspective time', on the one hand, and the scientific time-concepts, on the other, is not unconnected with another important issue – the mind-body problem.

It is a pleasure to express my thanks to the Trustees of the Leverhulme Fund for the award of an Emeritus Fellowship during part of the period when these studies were carried out. I am also deeply indebted to Dr. Michael Redhead for reading the original manuscript and for very helpful discussions, and to Professor Heinz Post for allowing me to enjoy the good and stimulating company of his department at Chelsea College.

University of London
Spring, 1981

Kenneth Denbigh

Table of Contents

Part I

Time as a Many-Tiered Construct

Chapter 1

The Problem Situation

§ 1. Introduction. The concept of time is not one we could easily do without and yet it presents us at almost every turn with tantalising paradoxes and largely unresolved issues. No doubt it was first created by the ancients to enable them to cope with the fact that *things are changing;* the clouds are moving and changing their shapes; plants are growing and withering; the positions of the heavenly bodies are slowly shifting; and men themselves progress inevitably from birth to death. The great value of the time concept is that it provides a systematization; all such events and processes of change can be treated as elements within a unique serial order.

Yet we quickly find ourselves faced with a peculiar question: Does time itself change? For certainly we speak of one day or of one season changing into another. Do we therefore require a *second* sort of time with which to describe the supposed change of time? But, if so, the second sort would need a third, the third would need a fourth and so on. The supposition that 'time' changes leads immediately into an infinite regress.

Another familiar difficulty concerns 'the present'. In an important sense I am always *in* my present; I am confined to it and can experience no other 'parts' of time. Yet we speak confidently as if there were other parts, those which we call 'past' and 'future', even though these appear in some sense not to exist; only the present is real. How strange this is! The present has the character of a knife-edge, a passing moment which separates a non-existent past from a non-existent future. How, we may well ask, can reality manifest itself so fleetingly and then become nothingness again?

'Time' was thus aptly described by C. D. Broad as "this most perplexing subject", and this leads one to enquire: What are the basic sources of this perplexity?

One of them, I suggest, is that time is often mistakenly thought of as if it were some sort of *existent* – indeed as a highly honorific existent, one which deserves a capital T as it were. This reification of time needs to be avoided. 'Time' is not 'out there' as a substantial thing like a river in flow; it is rather an abstract entity, *a construction.* The things which *are* 'out there', on which the construction of time is based, are the material objects and their events and processes. Before predicating 'existence' of some entity one of the questions to be asked is: Does it exist in time? and it would be peculiar, to say the least, to claim that time exists in time. 'Time' is not an existent.

A second source of perplexity is that 'time' is by no means a unitary concept. As has been said, the notion of time is required for dealing with the phenomena of change; yet there are many different kinds of change and, depending on the particular kind on which attention is focussed, *several alternative time concepts* are arrived at. Perplexity arises when the attributes of 'time' possessed by any one of these concepts are transferred indiscriminately to the others.

In this book I shall be concerned with three of these distinctive concepts: time as it is to conscious awareness; time as it is to theoretical physics; and thirdly time as it is to thermodynamics and to the evolutionary sciences such as biology. It is of importance to enquire about how they are arrived at and about the attributes of 'time' which they present; and further whether it can be said that any one of them is 'more true' than the others (and should be regarded as *the* concept of time) or whether they are interdependent and form some sort of hierarchy.

Towards making a beginning consider a typical utterance which exemplifies certain aspects of time-as-it-is-to-conscious-awareness: "This morning I was at a meeting and now I am writing". This involves the notion of a reference moment, the 'now' or 'present', and it also involves the notion of earlier states of affairs, those which are in the 'past' and are no longer real for me now. No doubt I could have expressed myself differently and in a tenseless form, not requiring the reference moment, by quoting the times and dates of my meeting and of my writing. But that would have been unnatural; it is not how we actually speak. The sense of an ongoing, allied with the awareness of a transient state of affairs (the 'now' or 'present'), is of the very essence of time-as-experienced. If it were to be asserted, on the contrary, that the ongoing is illusory, this would be paradoxial since even to think such a thing 'takes time'. All activity of the mind is a succession and thus the thinking of the proposition "succession is an illusion" is itself a succession and thus appears to be self-falsifying.

Yet theoretical physics offers a very different picture. Physics as such (as distinct from physicists in their daily lives) treats all 'parts' of time as if they were on a par with each other and it knows of no means of picking out a unique moment, the now or present. The t-coordinate is an undifferentiated continuum, and, if this coordinate is 'taken for real' as has been the tendency among many scientists and philosophers, the familiar distinction between past, present and future, so important in human affairs, comes to be regarded as a mere peculiarity of consciousness. It is as if every event along the coordinate is, in some sense, 'equally real', even those events which (to us) 'have not yet happened'. On this view of the matter it is a function of consciousness that we 'come across' those events, experiencing the formality, as it has been said, of the events 'taking place'.

Another important feature of the time of physics is that, as far as theory is concerned, *time has no intrinsic direction.* Here again physical time differs greatly from time-as-experienced. The basic theories, quantum theory and relativity, treat all event sequences as being reversible and therefore the theories in themselves, which is to say apart from the imposition of arbitrary 'initial' conditions, offer no distinction between 'time forward' and 'time backward'.

If this isotropy of the t-coordinate is taken literally, it would appear that the various natural processes, processes such as entropy increase and the apparent recession of the galaxies, are nothing more than temporary occurrences – temporary, that is to say, in relation to immensely vast durations of time. The resultant picture is that of a universe which, in some sense, is 'going through' endless repetitive cycles. Thus a hypothetical film-strip of cosmic 'evolution' would make equal sense whether it were projected in the one direction or in the reverse.

The third time-concept, that of thermodynamics and the biological sciences, has an intermediate character. It resembles the time-concept of theoretical physics in having no unique moment, the now or present. Yet it is similar to the 'time' of conscious awareness in so far as it regards the sequence of events as irreversible; any two events which are related as *earlier than* and *later than* cannot in general be interchanged with each other. If the one event is earlier than the other it cannot, on another occasion, be made the later one – not, at least, when the events in question are *total* – i.e. as including all consequential changes in the environment.

Here we have the familiar basis of the Second Law and of course the notion of irreversibility plays an almost equally important role in biology – even if biological processes, as I believe, cannot be fully characterised by entropy change. Sufficient for the moment to say that it would seem entirely false to suppose that either an individual life or the evolution of a species could ever proceed in the reverse sense. Ontogeny and phylogeny appear as being 'one-way only'.

Yet another aspect of irreversible change is displayed in psychology where it concerns the processes of perception and cognition. Once we have seen or known something we can never 'unsee' or 'unknow' it; we can play the role of observers but never that of *un*observers. This being the case the question arises whether it is self-consistent, in an epistemological sense, to use a concept of time, as in theoretical physics, which does not recognise this important limitation on the mental processes by which physics is created.

This leads on to another way of expressing my basic problem and this is the question whether 'time' belongs more to the external world or to the mind. On the face of it time is quite menacingly real and external; there is nothing whatever we can do to prevent its 'passing' or our own ageing; time presses itself inescapably upon us. Yet that is to talk in the style of reification. This apparent externality or objectivity of time must surely not go unquestioned. As has been said, we are dealing here with a conceptual entity, an abstraction, and one which depends on several more primitive terms such as 'earlier than' and 'later than' and 'now'. It will need to be examined whether or not the essential meaning of these primitive terms presupposes acts of human judgment.

Eddington (1935) remarked: "In any attempt to bridge the domains of experience belonging to the spiritual and physical sides of our nature, time occupies the key position." A clearly contrary attitude was expressed by Reichenbach (1956): "There is no other way", he said, "to solve the problem of time than the way through physics." These, I believe, are points of view which deserve very careful study.

§ 2. Constancy and Change. The problem situation I am dealing with is also concerned with the relative significance of constancy and change respectively. Here again there is a very big divergence between the world-views of physics and the life sciences and this is a divergence which amounts almost to a contradiction.

As is well known, physics (apart from thermodynamics) adopts an Eleatic view for it tends to regard only those things which are constant and changeless as providing an adequate basis for the understanding of nature. Hence the continued search for the 'fundamental' particles, those which are the true 'building stones'. Hence also the great importance in physical science of the various conservation principles, associated with the notions of determinism and of symmetry. What is called 'change' is seen as nothing more than a transient alteration in the spatial distribution of fundamental particles, or of the various conserved quantities such as energy or electric charge. Furthermore, when these rearrangements are regarded as taking place deterministically, the world comes to be viewed as being passive and inert, and as having no initiating powers of its own. If determinism is taken literally, *nothing essentially new ever comes into existence.*

This is paradoxical as well as peculiar. Causal influences must here be regarded as being merely *transmitted* – they are transmitted from earlier and yet earlier states of affairs and thus go back to an inscrutable Beginning, or to an infinite past. In short there is an endless causal regress since 'causes', on this view of the matter, can never initiate themselves and never make their appearance for the first time.

Very different indeed is the underlying philosophy of the life sciences whose central interest is not the constancy of organisms (there is no constancy) but rather their processes of change; for instance the development of a creature from an egg cell or the evolution of a whole species. What is particularly significant is that the concept of evolution, so basic to biology, not only gives emphasis to 'one-way' temporal development but also to the appearance of genuine novelty. Biology, unlike physics, sees nature as having a quality of creativeness or of inventiveness. During the 'course of time' radically new things come into existence, and among these are life and consciousness and mind.

To be sure it is just as useful in biology as in physical science to assume the existence of entities which are relatively permanent and unchanging, and which thus provide a continuing substratum *. For example, the

* Notice that whenever we speak of 'change' we need to be able to identify a 'something' which undergoes that change. Thus the notion of change always implies the existence of objects which have a sufficient degree of unchangingness to ensure their self-identity.

genes were conjectured to exist before there was any experimental evidence for them; they already had an important value as 'theoretical entities', namely to provide an explanation of the stability of inheritance and, at the same time, of its variations. In short, the genes are to biology what the atoms are to chemistry.

Even so, the life sciences are influenced by a paradigm of a very different sort – by the notion of *causal efficacy*. There is good reason for this; unlike the entities which are studied in physical science, living things present themselves as if they have a certain innate power, the power of *initiating* action, the self-moving power which is related to the human sense of free-will.

Although the notion of causal efficacy is metaphysical, it is perhaps no more so than are several other heuristic ideas which influence the development of science. Aesthetics and ethics inevitably enter in. For instance (referring back to an earlier point) it might be said with some justice that the interest of the theoretical physicist in a directionless time coordinate arises from his aesthetic desire for perfect symmetry. But equally it might be said that those who oppose that view of time are influenced by ethical considerations – by the belief that the universe must have some 'point', which thus brings in the temporally asymmetric notion of purpose. Perhaps it is anthropocentric to adopt such a belief; perhaps we should aim instead at conceiving the world in a kind of timeless sense, without purpose or desire, *sub specie aeternitatis*. Yet this view too comes from outside of science and is metaphysical.

Chapters 2 and 3 will describe in outline how the time concept is constructed on the basis of more primitive temporal notions such as 'the present' and 'earlier than' and 'later than'. This will result in the first place in a concept of time which is objective in the broader sense of being capable of public agreement and it will result, secondly, in the concept of the t-coordinate ('scientific time') which is objective in the stronger and more restricted sense of being independent of man's own existence.

What appears as a paradoxical consequence of the development of 'scientific time' is that 'the present' disappears completely, even though it played such a vital part in the original construction. Chapter 4 will therefore be concerned with the question whether or not 'the present' is indeed objective in the restricted or scientific sense. In dealing with this problem it will be contended that 'time' should be regarded as *a many-tiered* concept. For the special, but limited, purposes of physical theory, as applied to elementary phenomena, a sort of minimal or pared-down concept of time is sufficient. Yet this is quite inadequate for the purposes of describing more complex phenomena. Once it has been accepted that consciousness and mind are fully objective processes within the world, it must also be accepted that a notion of time is required which is much richer than it is to physics.

§ 3. Natural Processes. In Part II I shall turn to the various natural processes, the temporal ongoings, and these may be divided into two distinct classes, the dissipative and formative classes respectively. What I have called the dissipative class includes the thermodynamic (entropy) creating) processes, together with the expansion of wave fronts and the presumed expansion of the universe as a whole. Although all of these processes seem to indicate the reality of a 'one-way' ongoing, it has remained a problem ever since Boltzmann how processes of this kind can be made fully understandable within the context of a body of theory which is based on idealized reversible motions and t-invariant laws. Thus it remains controversial whether the supposed 'arrow of time' is fundamental or whether it appertains only to a temporary epoch. In an important sense this is a matter of whether we attribute a greater degree of significance to the observed physical *processes,* or to the t-invariant physical *theories.* For it might be said, on the one hand, that the t-invariant theories do not fully account for the observed processes – they have to be supplemented by additional information provided by contingent boundary (or 'initial') conditions. On the other hand, it might contrarily be claimed that the t-*non*-invariance of many physical processes is only apparent, and that, in the very long term, these processes would display the complete temporal symmetry required by the theories.

Before arriving at the formative processes, I need to take up a position on the age-old issue of determinism versus indeterminism, and this is done in Chapter 5. I shall argue that science offers no empirical support for the thesis of *ontological* determinism – i.e. the doctrine that events really are, in some mysterious sense, rigidly determined in advance. That particular version of determinism, making an absolute claim, appears incapable of being either corroberated or refuted. On the other hand, *epistemological* (or predictive) determinism is a reasonable thesis and is not absolute. It is a matter of degree. Nature seems usually to present us with a *mixed* situation, one in which there are simultaneously a number of factors tending towards a predictable outcome of a later event, and a number of other factors, chance factors, which do not. In classical science chance was held in rather poor esteem; it was 'mere' chance since it militated against the expectation that there are 'laws of nature' which are absolute. Due to the modern realization that there are perfectly good scientific laws which are 'obeyed' only in a statistical sense, the way has become open for an upgrading of the notion of chance and this, I believe, is important for the understanding of the life processes.

Previous concepts of 'emergence' have often been rightly criticised on the grounds that they imply discontinuities of a kind which are objectionable in science. In my own treatment, in Chapter 7, I shall point out that so long as there is an element of chance, however small, in the occurrence of events this will allow of genuinely new things appearing 'in the course of time' without the need for supposing that new principles, new laws, suddenly take over at various levels. If so, the natural order can be regarded as having a self-creative or inventive character. This is strongly exemplified in biology where the opportunities offered to the organism by chance events are reinforced by powerful mechanisms of selection and amplification.

What may be called the class of formative processes are those which may be loosely described as 'building up' activities – primarily the activities of living things. In my view it would be wrong to seek to characterize them in terms of increasing 'order', as has often been proposed. The concepts of 'order' and of 'organization' need to be sharply distinguished and if this is done the formative processes are seen to be in no way contrary to thermodynamics. But they do seem to indicate the existence of a non-conserved function which is entirely independent of entropy. This point is tentatively developed in Chapter 7.

In the final chapter, where I deal with consciousness as a temporal process, it will be suggested that the psychological 'arrow of time', the sense of a unidirectional ongoing, is probably related to the irreversible character of mental processes. Furthermore that the 'unity of consciousness' is dependent on the specious present which, because it has a finite duration, has the capacity of holding together as a structured whole what would otherwise appear as merely atomic perceptions and cognitions. This leads on to a consideration of creativeness in mental activity and this can perhaps best be understood by supposing that the brain, particularly the human brain, is the locus of an immense profusion of chance events, occurring in the brain structure and possibly at the quantum level, together with a correspondingly vast number of acts of selection and amplification.

Although this chapter is written from an entirely naturalistic viewpoint, it will be held that it is proper for naturalism to adopt *higher level concepts*, concepts which are appropriate to man's own level, as well as the lower level concepts which are appropriate to physics and chemistry. That would be in no way contrary to reductionism when this is regarded as no more than a useful methodological principle. The continuity of nature needs to be seen as being a continuity from above as well as from below, since otherwise there is no place for certain logical concepts which are essential to science yet are in no way derivable from it. In my view it is thus not at all inconsistent to regard the brain events as being *physical* and yet to think of the brain's superb selective system as operating according to *mentalistic* criteria.

To summarize, my aim is to deal with the 'time problem' in a manner which does not overlook the profoundly temporal character of consciousness and which may also go some way towards removing the divergence between the attitudes to time of theoretical physics, thermodynamics and the life sciences respectively.

Chapter 2

The Objectivity$_1$ of Time

§ 1. Introduction. It will be useful to begin with a brief passage from *The Critique of Pure Reason.* At the commencement of his discussion of 'time' Kant says: "Time is not an empirical concept that has been derived from any experience. For neither coexistence nor succession would ever come within our perception if the representation of time were not presupposed as underlying them *a priori.* Only on the presupposition of time can we represent to ourselves a number of things as existing at one and the same time (simultaneously) or at different times (successively)." ... "... the concept of alteration, and with it the concept of motion, as alteration of place, is possible only through and in the representation of time; ..."

One wonders what Piaget would have thought about this. Kant seems to be saying that we understand 'time' in advance of being able to use such words as 'now' and 'later than', and also in advance of having any idea of change and motion. He considers 'time' as it is already understood by the adult – and by quite an intellectual adult at that. To be sure it may be that Zeus knows 'time' in Kant's kind of way but in humans the notion does not spring fully armed out of the head. This indeed is what Piaget (1946, 1952) discovered in his studies on the development of temporal concepts in the child. The abstract notion of 'time' gives considerable difficulty to the young child and he or she is not able to use it properly until the usage of much simpler and more concrete temporal words has been mastered. Piaget also found that an adequate understanding of motion and of velocity actually *precedes* the child's comparable state of understanding of time.

In spite of Piaget one finds that a number of modern writers adopt much the same viewpoint as Kant. For instance Reichenbach (1957), after a very short preamble, proceeds to describe in great detail *how time is measured.* He does not tell us *what* is being measured; he does not inform us what there is in common between, say, 'June 16th' and 'December 10th' which constitute them as members of the class of 'temporal intervals'. For Reichenbach, as for Kant, this question seems not to arise. Like most other writers on the subject they take 'time' as being already well-understood.

The first point of my own I want to make is that this is a mistaken view. The notion of 'time' is not of sufficient clarity and simplicity for it to be regarded as a primitive term. For if it were indeed an adequate primitive why should there be such continuing controversy concerning many of its attributes? For instance concerning whether or not time is infinite in extension, whether or not it is continuous, whether or not it has an intrinsic direction, and whether or not 'the present' is objective. It is precisely because 'time' has this problematic character, involving these largely unanswered questions, that it cannot be regarded as an elementary notion. Far better to regard it as *a theoretical construction,* one which is built up on the basis of primitive terms which are much more clearly and directly understandable than is 'time' itself.

This leads to the 'relational view' of time as accepted throughout this book. It is the view that time is a relationship between events, and that without events there could be no concept of time. * But before proceeding further on these lines it will be useful to distinguish between two commonly used meanings of 'objectivity':

Objectivity$_1$. This relates to statements which can be *publicly agreed.* In a temporal context it would include, for example, "Yesterday it was raining" and "The clock now reads 9.15".

Objectivity$_2$. This relates to statements about things which can be held to exist, or about events which can be held to occur, quite independently of man's thoughts or emotions, or of his presence in the world (except where in the latter case they refer to his body.) For instance a proposition to the effect that the Cambrian rocks were formed before the Devonian would refer to a supposedly objective$_2$ event occurring previous to man's own existence.

Of course these two kinds of objectivity are not mutually exclusive; most of the objective$_2$ statements can be publicly agreed and at least some of the objective$_1$ statements are also objective$_2$. Nevertheless the second kind is obviously a stronger and more restrictive form than is the first and is what is usually meant by 'objectivity' in science.

Now it is a truism that we have no sensory apparatus which provides knowledge of temporal relationships as we do have senses which provide knowledge of spatial relationships. Our experience is always within the 'now' or 'present' and even our memories are present phenomena. Thus when I assert "Yesterday it was raining" this depends on a *present* memory (or a present record of some other sort) of what happened yesterday. The

* This is a philosophical, far more than a scientific, issue and one which I am not well qualified to argue. But it may be added that my own views on the matter are close to what Newton-Smith (1980) has recently called methodological reductionism.

fact that such a statement can nevertheless be publicly agreed indicates that people who were in my vicinity yesterday formed *the same* memory – and of course science understands this memory in terms of physical imprints in those people's brains and mine. The imprints are thus regarded as existing objectively$_2$, and this helps to provide a rationale for the correspondence between objective$_1$ temporal statements and objective$_2$ temporal statements.

The scientist, I believe, should not regard objectivity$_1$ dismissively! Indeed concerning 'time' a very adequate concept was created, on the basis of publicly agreed experience, long before there was any real science. It had been clearly realized by the ancients – although perhaps never explicitly stated – that there is a unique serial order of events (i.e. that 'time' is one-dimensional) and this, of course, was the foundation for the setting up of a trustworthy system for the dating of events by the use of calendars and primitive clocks. The objective$_2$ concept of time, to be discussed in the next chapter, was thus a late development. One of its origins was the theory of mechanics; another was the understanding of geology and of Darwinian theory which allowed it to be seen that the temporal order exists quite independently of man's own existence. Having achieved the objective$_2$ concept it then becomes reasonable, in its light, to reconsider the objective$_1$ temporal notions, such as simultaneity, which are at the roots.

This point should perhaps be made in a little more detail since it may appear to involve circularity. Suppose one makes the *initial* assumption that it is the thinking entities (such as myself and my reader) which constitute what is primarily real. By using the resources of science the thinking entities proceed to construct theories about what they perceive. An important feature of the theories is that they posit an *external world*, one which is assumed to exist independently of the thinking entities – i.e. it is objective$_2$. It is then found to be a reasonable deduction from the theories that the thinking entities are themselves the products of physical processes taking place objectively$_2$ in that external world. The thinking entities, it would appear, have descended, over evolutionary periods of time, from very primitive ancestors which, in their turn, arose naturalistically out of inanimate material. In short the effect of the theories is that the existence of the thinking entities *can be accounted for*. We thus correct the initial assumption by now regarding the objective$_2$ world as the primary reality. And this is not a circular argument. It is simply that one has passed, via the strength of the scientific account, from an idealist to a realist conception.

Such indeed is my own outlook. Even so I believe one cannot adequately study 'time' without considering its foundations as objective$_1$ since it is here, in publicly agreed experience, that the tacit understanding of the primitive terms is to be found. That is what the present chapter is about – and I shall not include clearly subjective features such as the estimation of temporal duration and alleged precognition, which are dealt with in books on the psychology of time.

§ 2. A Minimum Vocabulary of Temporal Words. The discussion of 'time' has been much bedevilled by the use of slipshod or pictorial language ('time's flow', etc.) and it therefore behoves us to choose our basic temporal words very carefully. There are many pitfalls.

Suppose we were to begin by saying that temporal experience presents itself as an ongoing or as a succession; that would be to assume that we have a prior understanding of 'ongoing' or of 'succession' and indeed of 'presents itself'. For these are words which already (already!) have a temporal connotation and which therefore depend on temporal presuppositions.* Perhaps we might start afresh by saying that not all 'events' 'occur' together, 'simultaneously', and that many of them 'occur' one 'after' another. Yet here again we are using words which may well presuppose precisely the conclusions we hope to reach. So too if we say that some of our experiences are 'past', or are 'no longer the case', whilst others are 'occurring now'. And the same problem arises when we use tenses, as when it is said that certain events have occurred, that others are occurring and that still other events may occur.

Evidently the very words we use to express our ideas about time already contain temporal presuppositions. This seems unavoidable and yet, if we are to avoid circular arguments, we need to be aware that these presuppositions exist, deeply embedded in language. A number of philosophers, such as Quine, Goodman and J.J.C. Smart, have advocated the use of a tenseless form of speech as a means of reducing temporal commitments. It certainly helps, but doesn't achieve everything that might be hoped for. For instance suppose it were said: "A light is shining in the house at 9.15 p.m., October 4, 1891". This utterance is tenseless, in the way these philosophers recommend, and it may be a true statement about some recorded event. Nevertheless it presupposes *a direction* of time since it assumes that the 'light' is, in fact, an emitter and not an absorber (Denbigh (1953, 1972); it assumes that the 'light' and its energy 'source' are losing energy, and not gaining it, in the (presumed) forward direction of time. The same presupposition is made if one speaks, say, of what 'happens' when hot and cold bodies are put into contact, for this is to suppose that indeed there 'has been' a 'putting together' and not a 'taking apart'. It is present too whenever we say that we 'observe' an event, for this is to preclude the converse process, if it is imaginable, of *un*observing.

Since, as has been said, 'time' itself is not a primitive, it will be useful to attempt a rough classification of temporal words before deciding on the best choice of primitives. The following list groups together various terms which are closely related:

(a) present (now); past; future
(b) earlier than (before); simultaneous with; later than (after); during
(c) event; occurrence; happening; process; the verbs 'to occur', 'to happen', 'to take place'

---

* In the present section many of these words will be picked out with single inverted commas in order to draw attention to presuppositions which might otherwise be overlooked.

(d) temporal interval (duration; period; time lapse); moment; instant
(e) temporal order; sequence; succession; transience
(f) the verb 'to be' and its tenses
(g) the verb 'to change', together with verbs and nouns denoting specific sorts of change
(h) the verb 'to become'.

Closely synonymous words are shown in brackets. Even so there is seldom complete identity of meaning; for instance 'the present' can usually be used almost interchangeably with 'now', but if it were said "World War II was present" this could not be replaced by "World War II was now". In the former sentence the temporal indication is provided by the verb since it is a *past* present which is referred to (Goodman 1966:365).

A number of authors have argued that the most suitable primitives are 'later than' and 'event', together perhaps with the verb 'to occur'. * However, in my view, that particular choice of primitives is insufficient. It is particularly insufficient as it applies to the construction of time as objective$_1$ (and, as has been seen, a tacit understanding of objective$_1$ time precedes the construction of time as objective$_2$). The publicly agreed temporal order depends quite fundamentally on the notion of 'the present' and this cannot be disregarded. Especially important is the sense of events 'being present' to us. Thus we cannot say of an event B that it is later than an event A unless we are aware of both events as being, or as having been, *presented* to us. It is only when this condition is satisfied that events can be placed in an order according to 'later than'. The notion of presentness logically precedes any other temporal concept.

Let me put this important point a little differently. It is true that the temporal order (whose construction will be outlined in § 4) is an order of events, or more particularly of the momentary states of things. Yet it needs to be asked: Which particular momentary states do we pick out? For obviously we don't group together *any* state of one entity with *any* state of another; we conjoin only those particular momentary states of two or more entities which appear as *co-existing within the same present* – and fortunately this is a matter which can be intersubjectively agreed. Thus the notion of 'that which is', or of co-existence, or of simultaneity with a person's present, is an even more essential and primitive notion than is 'later than'.

This point, which has been overlooked by many authors, is given some further support by the way in which we place past events in an order, especially those past events between which there is no causal connection. For instance I don't directly cognise my visit to York as having been either earlier than or later than the earthquake in China. Rather do I have the impression of the one event or the other as having the greater degree of pastness relative to my present. The judgment involves a triadic relation in which 'the present' plays the part of a reference state.

---

* Quine (1976) uses only 'later than' since 'event', he says, can be defined as its referent.

In short my view is that 'presentness' is an essential feature of time as objective$_1$, even though it has no function, or seems to have no function, in time as objective$_2$. But of course it would not be true to say that it is the only primitive term. Far from it. With 'time' we are concerned with several basic ideas and these have to be taken *together* and not by a stepwise logical process. A similar point has been made by Joske (1967:30) in connection with the concept of 'material object': "It is only a slight overstatement of the case", he remarks, "to say that the possibility of our saying any of the things we can say about material objects depends upon the possibility of our saying all the other things."

Nevertheless it will be salutary to aim at expressing what can be said about 'time' by means of a fairly restricted vocabulary, one which avoids certain locutions such as 'already', 'not yet', etc. whose temporal connotation, if it went unnoticed, might easily lead to a *petitio principii.* Accordingly my *minimum vocabulary,* at least for the purposes of Part I, will be as follows:

> The present (or now); earlier than, simultaneous with and later than; during; event; duration; temporal interval; moment; instant; temporal order; transience; the verbs to be, to occur, to change, together with verbs denoting specific sorts of change, and the tenses of these verbs.

For the sake of brevity 'past' (=time earlier than the present) and 'future' (=time later than the present) will also be used, together with all three of the terms 'earlier than', 'simultaneous with' and 'later than', even though there is a certain redundancy here. 'Moment' will be used as it is by Bergmann (1959:230) for denoting the specious present of conscious awareness whereas the word 'instant' will be reserved for a strictly zero temporal interval. A choice of primitive terms from this list will be made in § 4.

Perhaps a few words should be said about the verb 'to become' since this has been the subject of a good deal of controversy. In some senses it means only that a change is occurring – for instance "The leaves are becoming red". In another sense it seems to convey a deep ontological significance as meaning a 'coming into being' of physical objects – an appearance of what is real out of a non-existent future, a precipitation of being out of non-being. That, I believe, is a substantial issue and it is connected with the question whether or not 'the present' is objective$_2$ which will be taken up in Chapter 4. What, to me, is a non-issue is the spurious application of the notion of 'becoming' to *events,* or even to 'time' itself, as distinct from its application to physical objects. Much mystification arises if it is said that an event *becomes* present and then becomes past. As Wilfrid Sellars (1962:522, 556) has pointed out, in the framework of a 'thing' ontology it is *things* (and not events) which come to be and which cease to be; the event which is the coming to be, or the ceasing to be, of a thing is not a coming to be or a ceasing to be, but (like all events) is simply a taking place – i.e. it is an 'occurring' in the foregoing vocabulary.

Even more mystification arises if it is said that the future becomes present and then becomes past. McTaggart's (1927) paradoxes are partly the consequence of the *double* temporal reference which occurs in such assertions. Time is, of course, the reference variable for change and it is so constructed that all kinds of change of physical objects can be represented as the difference of some property (or properties) at different times. But time itself cannot be one of those properties. Thus it is fallacious to speak of the 'becoming' of the present, or of the 'passing' or of the 'changing' of time; what such phrases mean is nothing more than that a physical property, such as the position of the hands of a clock, has changed. As Goodman (1966:375) puts it: "... no time is at another time, ..."

§ 3. **Temporal Experience.** The foregoing paragraph, inserted at that point for convenience, has somewhat anticipated the abstract notion of 'time' which is to be arrived at in the following chapter. That is to say it has anticipated the notion of time *as a coordinate* – a coordinate which we are invited to think of as having 'places' within itself, places at which events can be located just as physical objects are given places in space.

It needs to be said however that our actual experience of time is utterly different from that of space and that it offers singularly little direct support to this conception of time as a coordinate. Before proceeding further along this path towards abstraction it will therefore be as well to pause and to remind ourselves of the real features of our temporal experience.

An important preliminary point is that under most circumstances we don't have any such experience! When we are attending to our affairs, just living, temporality is not at all noticeable; we don't even experience a present. Perhaps it might be said that we live as if 'from moment to moment', as other creatures are presumed to do. Yet even this phrase is too abstract since in our normal states of self-absorption we have no awareness of moments.

Thus it is only when we *self-consciously* examine our states of awareness that we become aware of experiencing a present state as being *that which is.* It seems doubtful that we should have this sort of awareness but for the existence of the faculty of memory which enables us to compare *that which is* with *that which was.* It is primarily this faculty which gives us the idea of a present which *is,* and of other presents which can be remembered, and each of them is a state of awareness of being aware.

Yet this awareness of a set of presents is quite different from the awareness of a set of physical objects as perceived by the senses. Being aware, by means of memory, of a set of presents is quite unlike being aware of a row of apples within any one present. For one thing because presents are not countable – they merge into each other. But more importantly because we cannot directly perceive events as having a 'time', as we can perceive objects as having a location. As was pointed out in § 1, it is an important truism that we have no sensory apparatus for giving knowledge of temporal relationships. The eye and the sense of touch provide an immediate awareness of spatial extension, but we have no senses which offer any knowledge of temporal extension – except perhaps for that very short duration which is the 'specious present'. We are always 'in' our nows or presents and thus we have no direct knowledge of anything other than is found in these presents. Even our memories and records are present phenomena; they exist *now* – as traces in our minds, as symbols in a manuscript or as fossils or strata in rock formations. Thus the construction of 'time' depends on inference and is largely based on the faculty of memory.

Consider the sense of transience in a little more detail. It continues to be extremely strong even when we have no perceptions of external events (or more accurately expressed, when these perceptions are so minimal as not to be conscious). Think for instance of lying in bed in a dark and soundless room; if we seek to remember our thoughts by consciously going over them again we find that they can be arranged in an order α, β, γ, etc., such that γ may be said to be later than β, and β may be said to be later than α, and so on. The thought train is thus a *series* such as will be considered in much greater detail in the following section. Furthermore we can proceed to a second stage where we go over our memories of our memories, and perhaps still further to a third stage – although at the expense of some mental gymnastics – where we have memories of memories of memories.

If memory of the past is thus one of the important factors in the construction of 'time', it is by no means the only one. Another such factor is anticipation. This is perhaps even more biologically useful than is a rudimentary memory since an anticipation of 'what is about to occur' is essential to the averting of danger.

Yet a further factor is the awareness of change 'within the present'. A familiar instance is our ability to observe directly that the second hand of a watch *is moving,* whereas of the minute and hour hands we can only report, by use of memory, that they *have moved.* Provided that a movement is neither too slow nor too fast we are able to perceive directly that one state of whatever is in motion is later than another state.

Hence arises the concept of the 'specious present' (or 'perceptual present'). To use William James' expression, it is a present which is not like a knife edge but is more like a saddle-back. It can perhaps best be understood in terms of a short-term memory having a particularly high intensity, and this may be relatable to be presence of chemical traces at the receptor organs (say at the retina) which do not fade instantaneously when the external stimulus has ended. Be this as it may, the biologically useful effect of the specious present is to give to temporal awareness a certain small degree of 'spread', analogous to the extensive spatial spread which is provided by the sense of vision. We can see many things 'at once'. If the analogous temporal spread were entirely lacking, which is to say if perception were truly instantaneous, we should presumably have a greatly impoverished sense of things being in motion. The specious present would thus appear to be as useful to other animals as it is to man; and indeed experimental psychologists have obtained comparable estimates of its duration (*ca.* 1/100 to 1/4 seconds) in other creatures. This will be referred to again in Chapter 4.

In summary it may be said that when we examine our temporal experience introspectively we find that its primary constituent is 'a present'. The faculty of memory then allows us to proceed to a slightly more abstract notion – that of one present after another. But of course this is not yet comprehensive enough to provide a concept of local time in which all presents, and more particularly all *events,* can be accommodated within *a single linear order.* To this I go forward in the following section.

Before doing so a little more may usefully be said about the transient character of the now or present. Although we are, in a sense, confined to this moment, nevertheless its actual content of perception is continually changing; an event which is (now) anticipated as a possible future event, may a little later be experienced as a (then) present event but only transiently for it subsequently takes on the peculiar status of a past event. The fact of transience thus gives rise to a familiar problem concerning truth values – Aristotle's problem about tomorrow's sea battle. Furthermore it is obviously not timelessly true to describe some particular event as 'being future' or as 'being present'; the truth value of descriptions in terms of futurity and presentness changes. For this reason it may appear preferable to make descriptions in terms of 'earlier than' and 'later than' since such descriptions may be said to be 'timelessly' true. As has been said already, many philosophers have advocated the use of tenseless forms of speech. *At all times,* they contend, Queen Anne's death *is* earlier than Victoria's. Such is the viewpoint of Goodman, Quine and J.J.C. Smart. On the other hand A. N. Prior, R. M. Gale and others have argued very cogently that descriptions in terms of past, present and future are not fully reducible to descriptions in terms of 'earlier than' and 'later than'. To attempt to do so entails a loss of information since, as far as human awareness is concerned, 'the present', is a very special reference moment.

And again in human experience there are important differences between 'past' and 'future' which would be suppressed if we sought to use only the temporal language of 'earlier than' and 'later than'. These differences are concerned with cognition and conation. The past, we feel, is capable of being fully known and in fact leaves traces or records. Thus I (now) remember that recently I lighted my pipe – and indeed have present evidence of it because I am now smoking. The future, by contrast, is not known and there are no (present) records of future events. I have no knowledge that I 'shall be' smoking when the clock shows 12.00 – or even that the clock *will* show 12.00. (Although surely it will.) On the other hand, in regard to conation, the past, it seems, cannot (now) be influenced whereas the future, we believe (if we are not rigid determinists), can be so influenced. And of course these distinctions are strongly reinforced by the impression that the past, in some sense, 'has already happened' whereas the future 'has not yet happened'.

Yet another familiar consequence of transience, and no doubt the most important from a scientific viewpoint, is that two temporal intervals can never be compared directly. We cannot lay a temporal tape measure along two temporal intervals, for purposes of comparison, as we can lay an ordinary tape measure along two spatial intervals. As has been said already, only one particular moment, the present, is accessible to us and in this respect 'time' seems to differ very significantly from 'space'.

The man in the street takes the transient aspect of things for granted; the scientist or philosopher finds it, in Broad's phrase, deeply perplexing. Why so? The influence of science, since Galileo and Newton, has been such that 'time' is treated as a coordinate and this results in those aspects of time – and transience is one of them – which are not spacelike appearing as strangely anomalous. Although one cannot know how future scientists and philosophers will view the matter, it can surely be said that the attempt to make time*lessly* true statements about temporal matters is paradoxical and must necessarily result in the loss of some of the essential aspects of time.

§ 4. The Temporal Order. Following this digression on temporal awareness I now return to the use of my minimum vocabulary for the purpose of presenting a fairly systematic account of time as objective$_1$. This will be done in outline only, and there will be logical weaknesses. As has been said, this chapter is concerned with the original foundations of the time concept and these were lacking in modern refinements, such as the notion of the continuum. Nevertheless what follows does, I think, provide a rationale for how time has been understood in past ages, and how it is still understood in most of its essentials.

From my vocabulary let us choose the following as the most primitive and necessary terms:

'present' (or 'now'), 'event'; 'later than'; together with the copula. Perhaps it should be said that 'events' will be taken as referring solely to events *as experienced* – i.e. to perceptions, and to things happening at the receptor organs, and *not* to their origins or causes which may be light-years away. The event-experiences are thus not subject to relativity effects. But of course they can be publicly agreed.

On this basis the theory of objective$_1$ time can be built up, with the help of some derived terms, as follows:

(a) There is a kind of event, to be called a *momentary event,* such that it can be experienced in its entirety within the same present. If two or more such events are experienced within the same present they are said to be *simultaneous with* each other. (It follows that it is necessary to distinguish between the set of events, many of which may be simultaneous with each other, and the set of presents of a particular person which may not.) If two momentary events are *not* experienced in the same present, one of them is *later* than the other. The latter is then said to be *earlier.*

(b) There is another kind of event, a non-momentary event or *process,* such that it consists of parts which cannot all be experienced in the same present. The latest present in which a process is experienced is called the *ending* of the process, and the earliest present in which it is experienced is called its *beginning*. (It should be added that the term 'latest' is related to 'later than' in the sense of there being no present in which the process is experienced which is later than the present in question. Similarly concerning 'earliest'). It will be clear that processes can *overlap* with each other; nevertheless each process can be regarded as consisting of very small parts, each one of them being capable of being experienced within a single present and thus qualifying as a momentary event.

(c) A *temporal interval* is a set of presents such that all of them are earlier than some particular present p_k and later than some other particular present p_i. Thus the interval defined by p_i and p_k is $\{p : p_k > p > p_i\}$ where > denotes 'later than'. Similarly the beginning and ending of a process enclose an interval which is called the *duration* of that process. We can now define *time* itself as the set of all temporal intervals, or more simply as the set of all presents. As such it is a rather ill-defined concept. No matter! – although the word 'time' is a useful one in certain contexts such as 'asking the time', for present purposes it is best to rely on the primitives.

(d) The relation 'later than' is asymmetric and transitive in the sense of the theory of relations. (This is an empirical statement, not an analytic one, for it is conceivable that events could be such that $e_2 > e_1$ and $e_3 > e_2$ does *not* imply $e_3 > e_1$. Indeed this would be the case if 'time' were a closed order, although there is no good evidence that it is – nor could it be *experienced* as such § 6.5.) Now it is familiar that a relation which is asymmetric and transitive is not sufficient to define a serial order – such as we expect the 'temporal order' to be – and the third property which is required is connexity. That is to say, if a relation is to be used for the purpose of establishing a serial order within a set of elements it must hold between *every pair* of elements of the set. Suppose we think in the first place of the set, *X*, of all experienced events, momentary or non-momentary. Within this particular set the relation 'later than' does not have the property of connexity since there can be pairs of elements of *X* – namely those which are wholly simultaneous with each other – such that neither the one nor the other member of the pair is later than the other.

Consider instead the set *Y* of my own (or any other person's) 'presents'. In order to avoid confusion between *the* present (the 'present present') and those other presents which are held in the memory, it will be convenient to use the word 'moment' as meaning any present. Now of *every* pair of moments, whatever the interval between them, it may be said that one member of the pair is later than the other. In the case of the set *Y* the relation 'later than' is thus asymmetric, transitive and connected and thus allows of all moments being placed in a serial order. But it should be added that there is an obvious weakness in this argument since moments, or specious presents, presumably merge into each other and this, of course, is one of the reasons why moments are replaced by 'instants' in the theory of time as objective$_2$.

(e) With the above an objective$_1$ temporal order has not yet been arrived at. As has been said the elements of the set Y are the presents of some particular person, P. These in themselves are not objective$_1$ since P's presents, as such, can never be directly identified with another person's. Thus although the presents provide a serial order for P they do not provide *a publicly agreed* serial order. To achieve the latter we must turn to the set X of all experienced events and proceed to partition it into subsets X', X'', etc., such that each subset contains those elements of X which can be perceived by all persons in their vicinity as being simultaneous with each other. In other words the 'presents' are used for the partitioning of X. * (Of course it must be assumed that the events in question are distinguishable from each other; the various persons must be able to refer to, say, a particular closing of a door and not merely to *any* closing. This particularisation – without which there is no unique serial order – is greatly facilitated as soon as we bring in the use of calendars and clocks whose readings may also be regarded as events.)

The subsets of X are thus so chosen that all momentary events within a particular subset are capable of being publicly agreed as occurring at the same moment. Clearly these subsets of X are in a 1 : 1 relation with the set Y of a particular person's presents and the set Y, as has been seen, is serially ordered by means of the relation 'later than'. It follows that the set X of experienced events can be partitioned into subsets of *momentary* events such that the momentary events are related to each other *either* by a symmetric relation of simultaneity within the subsets, *or* by the relation 'later than' as between one subset and another. What can now be called *the temporal order* is the order of the subsets.

(f) It seems that *all* experienced events can be accommodated within a *single* temporal order and this is a very remarkable empirical fact, one of the most striking simplicities in nature which have ever been discovered. There appear to be no events which cannot be so accomodated. In other words, *particular* sequences, such as the deaths of kings and queens, stages in the growth of a tree, the movements of the stars, etc., do not require separate temporal orders of their own. Furthermore the ordered set of events can be taken as including events which have been experienced by any number of generations of observers so long as these have been reliably recorded. And it can also be taken as including events such as the laying down of the rock strata whose occurrence is inferred rather than directly experienced.

To say that there is a single temporal order is intuitively the same as saying that 'time' is uni-dimensional **. It might not have been so. And indeed space cannot be so ordered for it is another empirical fact that the ordering of things in space requires *three* ordering relations, not just one.

---

* The reader who is familiar with Russell's early paper *On The Notion of Order* (1901) will recognise that I am using what he calls 'the third way' of creating a serial order.

** In modern mathematics 'dimensionality' is used in in a somewhat different sense. (Hurewicz and Wallman, 1948.)

The 'later than' relation would not be a sufficient ordering relation if time were not uni-dimensional.

As was said earlier, the realization that there is a unique temporal order was achieved long before the advent of science. It became more firmly established when methods of dating, based on periodic events in the heavens, had been devised and of course this process was carried a stage further with the invention of clocks. This *metricisation* of the temporal order will be discussed in the following section.

(g) Before doing so it should be said that a further empirical fact which has been tacitly used in the foregoing is that *we never experience exactly the same circumstances twice.* If *S* is a total state of the environment which is experienced on one occasion it is not experienced on a second. Even in *deja vu* our impression is that of some event or scene having been experienced *previously.* In other words the *deja vu* experience differs from the supposed original experience at least in this: that, however veridical it may seem, *it includes* the sense of there having been a precious experience and thus differs from that previous experience by the very fact that the latter did *not* carry with it the impression of *deja vu,* of the repetition of some yet earlier experience.

This non-recurrence of total states is closely bound up with the existence of irreversible phenomena. There are innumerable familiar instances of events, say the events *A, B,* and *C,* such that if they are experienced in the order *ABC* they cannot also be experienced in the order *CBA*. This, of course, was taken for granted long before there was any scientific analysis of irreversibility. It would have been regarded either as an illusion or as a miracle if the phases germination, flowering and decay in the lives of plants had ever been experienced in reverse. Similarly in regard to the raising of Lazarus from the dead.

The fact that exactly the same circumstances do not recur – or at least do not recur within the supposed history of the universe – is implicit in the use of 'later than' as the ordering relation for the temporal order. If this were not the case, and if it were possible to 'return' to exactly the same state (including the states of all records and memories), there would be no basis for claiming that this was indeed a 'return'. For there would then be no means for distinguishing, when all clocks and calendars read what they originally read, between the 'original' state and the supposed 'repeat' state. Presumably we should be dealing with 'closed time' – but with no means of *knowing* that we were. More will be said about the highly theoretical status of closed time in Chapter 6.

It will be seen that our experience of time differs importantly from our experience of space. Two points on a line can be experienced as interchanged, in regard to the ordering relation 'to the right of', by looking at the line from a different angle; but we have no means of experiencing an interchange of certain kinds of events in regard to the ordering relation 'later than'. Also it appears that there is much more to this difference than the mere fact that, because time is uni-dimensional, we cannot look at events externally, from 'a different angle' so to speak. Consider a one-dimensional space – a straight line. A one-dimensional being who lived in that line would presumably have no means whatsoever of telling the one direction along the line from the other. But we, who live in a 1D temporal order, do find that we can distinguish the one direction along the order from the other. The one direction appears to be *accessible* to us whereas the reverse direction is not. And quite apart from human experience, the existence of irreversible processes seems to allow for the two directions along the temporal order being *structurally* or *intrinsically* distinguishable; that is to say, they are distinguishable *from within the order* itself (Grünbaum 1973; 564). The same is true of the system of real numbers where the ordering relation is 'greater than'. * In other words the temporal order has something in common with the real numbers which it does not have in common with a dimension of space. This important point will be returned to, for dealing with more carefully, in connection with the thermodynamic view of time.

§ 5. The Measurement of Time. There is good reason for including this subject in the present chapter, rather than in the following one, since time measurement, at least in a crude form, long antedated the concept of the t-coordinate. The notions of the dating of events, and of the relative durations of different temporal intervals, had been fully assimilated within the objective$_1$ concept of time from an early period.

Quite a short section will suffice since the measurement of time, unlike much of the foregoing, has been amply discussed elsewhere. Indeed what follows is largely based on the presentations of the subject by Reichenbach (1957), Whitrow (1980) and Carnap (1966).

It has been seen that events at a particular location can be placed in a unique serial order, the temporal order. The problem we are now concerned with is that of providing a measure of separation within this order – a measure, that is to say, of temporal intervals. For this purpose, as Carnap points out, we need three things:

(1) We need to satisfy a *condition of additivity* **; if the ending of a temporal interval *a* is simultaneous with the beginning of another interval, *b*, we require that the total interval from the beginning of *a* to the ending of *b* is the arithmetic sum of the intervals *a* and *b* separately.

(2) We also require a *rule of congruence* – i.e. a rule which specifies when temporal intervals are equal. (3) Finally we need to choose a *unit of time.*

It has been realized for many centuries that periodic processes seem to offer the best candidates for the satisfying of these conditions. However

* Dr. M. L. G. Redhead has pointed out to me that the direction of the sequence of real numbers is not a purely ordinal one; it depends on the axioms relating to binary operations as well as on the axioms relating to order.

**As Whitrow (1980;219) remarks, additivity ensures that physical laws can be formulated so as to be independent of the particular times at which events occur; only the *differences* of the times of events need then be taken as significant.

from a logical point of view a circular argument arises here. It cannot be said that some particular recurring process is *regular* – i.e. that its periods mark out *equal* times – until we already have available some method of measuring those periods. And that method is precisely what we are looking for!

To make the point clear consider some examples of processes which recur: (a) the going up and down of a lift; (b) the beating of some person's heart; (c) the swinging of a pendulum; (d) the earth's rotation relative to a fixed star; (e) the characteristic internal oscillations of some particular type of atom. On what good grounds could it be asserted that the up and down trips of the lift keep 'less accurate time' than the heartbeats, or that the latter are less regular than are the processes (c), (d) or (e)? For lack of an independent means of measurement there are no such grounds. What in fact is done is to choose one of the latter types of process as providing the means of measurement *by convention.*

However, as Carnap points out, there is a significant reason for preferring any of (c), (d) or (e) to (a) or (b). Suppose that on some particular occasion there are 97 of Jones' heartbeats to every 100 of Brown's. This will not necessarily be the case on some other occasion, and the same non-reproducibility applies to other biological periodicities, and of course even more strongly to the motions of lifts. On the other hand it is an important empirical fact that if there are 97 swings of one pendulum to 100 swings of another, the same will indeed be the case on another occasion. Furthermore there is a similar reproducibility between, say, the swinging of a pendulum, the rotation of the earth and the oscillation of an atom. If 86,400 swings occur today during one rotation of the earth, the same will be found tomorrow – or very, very nearly. A similar characteristic of remaining 'in phase' with the earth's period of rotation applies to the atomic oscillations *. Furthermore the class of processes of the type of (c), (d) and (e) is very large indeed and also there appears, says Carnap, to be *only one such class.*

Therefore it is natural (but in no sense *necessary*) to choose any one of these processes as providing a unit of time. For instance if the unit 'the day' is chosen as the period of one rotation of the earth relative to a fixed star, temporal intervals would be measured as so many multiples or fractions of this unit. (The fractions of a day are conventionally chosen in terms of equal angles towards the star.) This measure of temporal intervals is then automatically additive in the required sense. During recent years the earth has of course been replaced by the caesium atom as the standard clock; the basic unit is no longer the day but is the period of a particular caesium spectral line.

* Notice however that the atomic oscillations are dependent on the electromagnetic force, whereas the periodic processes in the heavens are dependent on the gravitational force. If, as is now widely supposed, the gravitational 'constant' g has been slowly diminishing in value since the big-bang, the atomic and astronomical periodicities cannot be precisely in phase with each other and, in fact, define two basically different time scales.

It has still to be asked whether the unit can be said to remain 'unchanged' in some sense. Can it be assumed that a caesium oscillation marks out an equal interval today – an equal 'amount of time' – as it did yesterday? It would be attractive if this could be assumed, but it appears to be quite unprovable [c.f. Christensen (1976) and Roxburgh (1977)]. Obviously enough yesterday's temporal interval cannot be brought into coincidence with today's temporal interval, as we can bring two yardsticks into coincidence. And neither can yesterday's interval be 'measured' apart from its comparison yesterday with the chosen unit. Since all 'regular' recurrent processes remain in phase with each other, it cannot be assumed that they are not all 'changing' their periods *together* – i.e. in such a manner as maintains a constant phase relationship.

Perhaps it might be argued that it would be contrary to the Principle of Sufficient Reason to suppose that the periods of all these processes are 'changing' magically in harmony; no cause could be put forward why they should all 'change' together. Yet this is really beside the point. The significant point is that, for lack of a reasonable alternative time unit, the very notion of their supposed 'change' becomes meaningless. It is therefore entirely *a convention* that any one oscillation of the caesium atom defines an interval which is the same as that of any other oscillation of this atom (c.f. Whitrow 1980;44 and Newton-Smith 1980).

Further support for the view that the foregoing is a suitable convention is provided by the fact that isochronous intervals, as so defined, result in the laws of mechanics *taking on their simplest form.* As Reichenbach remarks, the laws do not *compel* us to believe that two successive periods of, say, the earth's rotation are equal since the laws are formulated on the supposition that they are indeed equal. But with this convention the laws, such as the Newtonian Law of Inertia, take on their well-known formulations. These are much simpler than would be obtained by supposing, say, that bodies falling towards the earth mark out equal temporal intervals in equal distances – and of course far, far simpler than would be obtained if biological rhythms or lift movements were taken as the basis. Admittedly the notion of 'simplicity' in science is by no means a simple one! Yet in this instance the ambiguities in using it are not very great.

We thus achieve a rationale for the measuring of time and, as Lucas (1973) has said, the fact that it is indeed a rationale is made evident by the way in which we regard our clocks as being *corrigible.* As is well-known, the earth's rotation is very slightly irregular relative to other time-keepers; so too are pendulum clocks due to small variations of g, and so on. Yet the laws of physics allow of small corrections being applied in a self-consistent manner. As was originally pointed out by Poincaré, time measurement is based on a consensus among many physical processes, together with their associated 'laws' and theories.

Before concluding it may be added that although the large class of periodic processes provide the accepted time measure, this class is by no means the only reasonable candidate for the purpose. A possible alternative would have been the class of processes of radioactive decay since all radioactive elements obey the same law – the Rutherford-Soddy law. This law indicates that the logarithm of the fraction of atoms which remain undecayed during a given temporal interval provides an additive measure of that interval. Furthermore the measure is the same (apart from a constant numerical factor which depends on the ratio of half-lives) for all radioactive elements.

Although this logarithmic measure is linearly related to the measure provided by the 'regular' periodic processes, it offers no advantages. Indeed it would be much less satisfactory in regard to precision. The Rutherford-Soddy law is a statistical law. It holds only 'on the average' since the actual number of atoms decaying during a given temporal interval is subject to random fluctuations.

So much concerning the measurement of temporal intervals. For dating purposes one also needs a reference time, taken as zero, and this too involves a convention. For obviously we don't know the exact 'time' when Jesus was born, not even to within a day. When we speak on a January 1st or 2nd of being in, say, the year 1978 what exactly do we mean? Paradoxical though it may seem, it is really to take *whatever is the present year* as being its own reference year! Thus when 1978 is the present year, what we are doing in effect on January 1st is to count *backwards* by 1978 units of a year to a time when Jesus was supposedly born. Presumably there is an official at Greenwich whose pleasant duty it is, each January 1st, meticulously to record the adding of 1 to the A.D., thus ensuring the soundness of the whole system of dating.

Chapter 3

The Objectivity$_2$ of Time

§ 1. Introduction. As has been said, the notion of the temporal order was pre-scientific and was based on publicly agreed experience. Nevertheless it had proved entirely adequate for the purpose of giving dates and times to events, and of course relativistic effects were still unknown.

There were however a number of weaknesses in the theory of Chapter 2 and one of them was its dependence on the notion of 'the present'. It will be shown in Chapter 4 that there is no strong evidence concerning whether or not 'the present' is objective$_2$ – i.e. whether or not it would make sense to speak of 'the present' if man did not exist. Furthermore the previous account depended on 'moments' – the duration of presents – and these are not 'instants'. All timing which is based on human awareness during a moment may well be subject to an error of about a tenth of a second.

Also from a logician's point of view it has seemed desirable to avoid propositions (if they can be called that) whose truth value changes. The statement "At present it is raining" may be true at one time but false at another. Thus if we are seeking to establish propositions which, like the propositions of mathematics, shall be timelessly true it is desirable, if it is possible *, to eliminate the token-reflexive or indicator words, such as 'the present' and 'now' and 'today', and also to eliminate the use of tensed verbs.

For reasons such as these the objective of the present chapter is a theory of time which provides a valid concept of events taking place in an ordered sequence even if man, with his awareness of 'the present', did not exist. What will emerge, of course, is the notion of time as *an undifferentiated t-coordinate,* one which makes no reference whatsoever to 'past', 'present' or 'future'.

Such a concept will clearly be much more abstract than is the concept of time as objective$_1$. It will be much more of *a theory* – and as such it could be wrong. Indeed it might be argued that it is bound to be unsatisfactory if only because 'the present' is the most primitive temporal idea. Yet that would not be a strong argument for it is a commonplace of science that

* As was said in § 2.3, a number of logicians regard it as not possible, or as being a mistaken aim.

what is basic in one theory need not be basic in another. The notion of 'force' was basic in the mechanics of Galileo and Newton but it played a much less significant part in the later mechanics of Hamilton and Lagrange. Analogously it may be said that the route by which we initially arrive at the concept of time need not necessarily be the same route as that which is adopted in a later and more abstract theory.

In a similar vein Quine (1976:236) has replied to the supposed objection that something tantamount to the use of indicator words is finally unavoidable, at least in the teaching of the terms which are to be made to supplant the indicator words. "But this", he says, "is no objection; all that matters is the *subsequent* avoidability of indicator words. All that matters is that it be possible in principle to couch science in a notation such that none of *its* sentences fluctuates between truth and falsity from utterance to utterance."

Very thorough discussions on the setting up of the t-coordinate are already available in books such as those of Whitrow (1980) and Grünbaum (1973). The present chapter will therefore be a commentary, at a more elementary level, on the main issues involved. Relativity will be deferred to § 6 and the intervening sections will be concerned with the issues: (1) Why are instants needed and how are they to be introduced?; (2) Shall the ordering relation be chosen as 'between', as in the so-called causal theory of time, or shall it be chosen as 'later than'? If the latter choice is made, what is to be the empirical criterion of one event being later than another if the criterion based on conscious awareness is to be avoided?

§ 2. Instants. Several reasons have been put forward why 'instants' are *useful* within the theory of time, but not all of these are good reasons for 'instants' being *necessary.*

It has been mentioned already in Chapter 2, § 4, that there is a weakness in using specious presents or 'moments' for the purpose of constructing the temporal order since these presumably merge into each other. This casts some doubt on whether the relation 'later than', as applied to the set of all moments, may be said truly to have the property of connexity. On the other hand if 'instants', each of strictly zero duration, can be introduced into the theory of time this difficulty would disappear. Within the set of all instants the relation 'later than' is not only asymmetric and transitive but it also holds quite clearly between *every* member of the set and this is the required property of connexity. Thus the assumption of instants helps to give logical rigour to the construction of time as a serial order.

A second reason for invoking instants is the desire for a comparable degree of rigour in the use of the calculus. Functions which are not continuous within an interval are not differentiable throughout that interval. Thus if 'time' were *not* regarded as being a continuum of instants this might appear to create difficulties concerning concepts such as velocity and acceleration. Yet useful though the calculus undoubtedly is to science, is it essential? Whitrow (1980;205), who is a believer in discontinuous time, has pointed out that finite difference equations can now be handled very easily by use of computers; the primary reason, he thinks, why physicists have clung to the hypothesis of instants has been their former mathematical convenience.

A similar comment may be made on a third argument in favour of instants and this concerns irrational numbers. The argument may be expressed by means of an illustration taken from Lucas (1973). Suppose it has been found experimentally that a body falling from rest traverses 64.4 feet in 2 seconds. This result is (minimally) sufficient to fix the value of g; thus if we now use the formula $s = gt^2/2$ for the purpose of calculating the time it takes for the body to fall half that distance we obtain the answer $\sqrt{2}$ seconds – an irrational number of seconds. The significant point is that the possibility of working out this result (and of course many similar instances could be quoted) would appear not to be available from the Newtonian theory if the temporal order were not representable by the continuum of the real numbers – that is to say by the 'gapless' sequence which consists of the irrational numbers as well as of the rationals. As Lucas remarks, it would be counter-intuitive *if there were no instant* at which the body passes the point at which it is 32.2 feet below its point of release.

On the face of it this is a strong argument for regarding time as being a continuum of instants. Yet is it not an aspect of the erroneous reification of time? It supposes that there has to be a sort of entity, an instant, which is 'at' $\sqrt{2}$ seconds. Furthermore we can never actually measure time (or distance) to the infinite degree of accuracy required to show that the time of fall through 32.2 feet is indeed an irrational number of seconds, rather than a rational number which differs only infinitesimally.

Suppose it were to be found, perhaps from some development in quantum theory, that time and space should preferably be regarded as being discrete. If that were to occur we can be sure that philosophy would quickly find the means of adapting itself! (Just as it has in regard to non-Euclidean space.) The turning up in calculations of irrational numbers would not be an obstacle, and neither would Zeno's paradoxes. Since the falling body really does traverse 32.2 feet, and since Achilles really does overtake the tortoise, philosophical analysis would find a way of coping with the situation, whether it be found from the side of science that 'time' is best regarded as continuous or quantised – or again, as yet another alternative, as being dense rather than continuous.

Perhaps it would have been as well if the noun-word 'time' had never been invented – for nouns always tempt us to believe that they denote something that exists physically. As defined in the previous chapter 'time' is the infinite set of all temporal intervals and, as such, its 'existence' is mathematical or logical, not physical. Thus, in my view, the question whether its smallest intervals are of finite duration or of strictly zero duration is the question concerning which of these two alternatives is the most self-consistent when the issue is considered in all relevant respects. The classical arguments have undoubtedly favoured the view that 'time' is best regarded as a continuum of instants. Yet further considerations may arise which will tip the balance in favour of the atomicity of time and the 'existence' of chronons.

But let's accept instants and proceed to enquire about the means by which they have been brought into the theory of the t-coordinate. Early in the present century Russell and others engaged in a strenuous attempt *to construct* instants from the phenomena of temporal experience. Finite temporal intervals can overlap with each other and it might seem that if one considered an ever-enlarged set of overlapping intervals the instant might be obtained as the limit of the ever-diminishing region of overlap. This led Russell (1914;124) to define an instant as "a group such that no event outside the group is simultaneous with all of them, but all the events inside the group are simultaneous with each other." Yet it did not follow that this definition necessarily resulted in a *continuum* of instants, a series which would be 'gapless'. A proof of this could not be obtained without adopting various assumptions. By the time of one of his later papers on the subject Russell himself (1936) had evidently become sceptical about the whole programme for he remarked that "the existence of instants requires hypotheses which there is no reason to suppose true – a fact which may not be without importance for physics." And he went on to say that if these hypotheses are not true, 'instants' can only be a logical ideal. Russell was perhaps hedging his bets in case physicists might eventually come to regard time as being discrete.

In view of these difficulties a number of authors, such as Grünbaum (1973) and Bunge (1968), have been content merely *to postulate* that time is a continuum of instants, rather than seeking actually to construct this continuum. The question remains an open one whether or not this postulate is in any sense *necessary* to the theory of time, as this is seen in the current state of philosophy. Without trying to answer this question beyond what has been said already, it is of interest to mention that Hamblin (1971, 1972) has developed a theory in which time is regarded as consisting of intervals rather than of instants. In its present form this theory is perhaps better adapted to the needs of logic than to those of science. Even so, its general method of approach fits in rather well with the views of those scientists who believe that a discrete model for time is to be preferred

The foregoing issues have been discussed from the side of science by Schwartz and Volk (1977). A comprehensive philosophical discussion, which includes a consideration of 'density' versus 'continuity', is given by Newton-Smith (1980).

§ 3. The Causal Theory and 'Betweenness'. Two methods have been proposed for establishing the objective$_2$ temporal order. One of them is the 'causal theory of time' and the other depends on the use of a criterion for 'later than' which is supposedly independent of human judgments.

The causal theory has had a long and tortuous history, from Leibniz and Kant onwards, but here I shall attempt nothing more than a brief outline of one of its modern versions, that of Grünbaum (1973) and van

Fraassen (1970). Earlier forms of the theory, such as had been put forward by Reichenbach (1957), were based on making a distinction between 'cause' and 'effect' but this was found to be defective. The distinction cannot be made without either: (a) already assuming in a covert form the very temporal order which it is desired to achieve, or (b) requiring the use of a 'mark' method. The 'marking' of an event is necessarily irreversible with the consequence that the theory is no longer purely causal but becomes dependent on thermodynamics. Thus it has nothing to offer beyond what is already available from thermodynamics concerning a criterion for 'later than'.

Grünbaum put forward a causal theory in which no distinction is made between 'cause' and 'effect' and which requires no invocation of irreversible phenomena. He speaks of events as being causally connected – or more particularly as being causally *connectible* (i.e. as having the possibility of being causally connected). Thus he is concerned with a *symmetric* relation and not with one, like 'greater than' or 'later than' which is asymmetric. This relation is taken as primitive. Furthermore he confines his discussion to macroscopic bodies and thus excludes consideration of the events of electrons, protons, etc. These macroscopic bodies possess the property of genidentity – i.e. of remaining *the same body* in spite of changes of their locations. The set of genidentical events which form a connected causal chain is assumed to occur deterministically and the chain of events is further assumed to have the cardinality of the continuum.

In short the essential notion in Grünbaum's theory is that a causal influence consists of a continuous sequence of events, each of them being fully determined. On this basis he obtains a definition of the 'temporal betweenness' of events but finds that this definition is incapable of distinguishing between *two alternative types* of temporal order. These are: (a) The open order whose spatial analogue is the infinite set of points on a straight line; (b) The closed order whose spatial analogue is the infinite set of points on a circle. In the latter case every point or instant is 'between' *any* pair of points or instants. Thus according to Grünbaum's causal theory, both 'open time' and 'closed time' are possible; neither can be ruled out, he says, except by reference to boundary conditions which are extraneous to the theory. Indeed he regards it as a failure of imagination if we do not seriously envision the possibility that 'time', in the very long term, may in fact be closed.

Leaving that issue aside and considering only 'open' time, what may appear as a puzzling aspect of the foregoing is how the relation of 'betweenness' can supposedly be used to create a serial order within a set of events, since that relation does not have the property of asymmetry. In fact the betweenness relation was not used by Grünbaum to create a serial order; it was used only to achieve a *coordinization* of events within a 'time' which was already assumed to be either linelike or circlelike. On the other hand, if the events were assumed to occur *in*deterministically, the most that could be asserted is a relationship of *statistical* betweenness, such as had been adopted by Reichenbach (1956;188). Where any event has probabilities of giving rise to two or more other events, the best that the causal theory can do is to establish a *causal net*, as distinct from linear causal chains, and it is not a property of the net that it has a serial order.

It will be seen that the motivation behind the causal theory is that of reducing spatio-temporal relations to causal relations. It would be out of place to attempt to summarise the various criticisms which have been advanced (Smart, 1971; Lacey, 1968; Sklar, 1974; Mackie, 1974), but many of them are concerned with this very point. It has been emphasised that the notion of spatio-temporal relations is much more deeply rooted than is the notion of causality, and therefore, if any reduction is to be attempted, it should be in the reverse sense to that which is attempted in the causal theory.

Furthermore (as pointed out to me by Dr. Redhead), the concepts of genidentity and of deterministic observables, on which the causal theory relies, have been undermined by quantum mechanics, and this is widely regarded as the most fundamental physical theory of our time. If so, it would appear that the causal theory is not in harmony with the mainstream of modern science.

§ 4. The Objective$_2$ Temporal Order as Based on 'Later than'. In this section I proceed to the second of the two methods referred to at the beginning of § 3. Discussion will be limited to events occurring at the same location and considerations arising from relativity will be deferred to § 6. Also it will be assumed that 'time' is 'open'. The universe's history, as far as is known, does indeed appear to be open. The formation of the galaxies was later than the 'big-bang'; the formation of the stars was later than the formation of the galaxies; and so on. Perhaps it cannot be excluded that the order of events is not open in the extremely long term and this question will be taken up in more detail in Chapter 6. Concerning the period of which we do have some knowledge we are justified in regarding the order of events, and thereby 'time' itself, as having the characteristics of an open order.

As has been seen the relation 'later than' as applied to instants has the three properties of asymmetry, transitivity and connexity which are required to give rise to a serial order. All this follows on, and more rigorously, from what has already been said about 'moments' in Chapter 2, § 4. But it is worth re-emphasizing that in human experience the two directions along the temporal order are intrinsically distinguishable – they are distinguishable from within the order itself. In the case of a straight line, on the other hand, the order is merely *extrinsic* since the imposition of the order by means of the relation 'to the right of' requires an external viewpoint (it is a *triadic* relation) and the order can be reversed by looking at the line from a different external viewpoint. It is the fact that the temporal order appears to have this *intrinsic* direction (which is something different from asserting that the one direction may be more *real* than the other) which is variously known as 'the direction of time' (Reichenbach), 'the anisotropy of time'

(Grünbaum) or more simply, but perhaps deceptively, as 'time's arrow' (Eddington).

The question now to be asked is whether an asymmetric and transitive dyadic ordering relation, equivalent to that which is provided by the human sense of 'later than', can be established in an objective$_2$ manner – i.e. independently of consciousness. It will be tacitly assumed in what follows that the sought for relation will also provide for an *intrinsic* distinction between time's two directions. Yet this is really a logically distinct issue and it will be taken up more fully in Chapter 6.

Now the concept of objectivity$_2$ does not require us to think of a world which is without scientific instruments and without readings of those instruments. (Otherwise nothing in science *could be* objective$_2$!) It requires only that the pointer readings be capable of expressing truths about the physical world even if man were not present to take the readings. Furthermore the instruments can be of a conceptual form. They can be supposed to have existed during the whole duration of the universe's history, and they can also be supposed to have been 'read' by means of ciné cameras which have been recording continuously, and are still recording.

What seems to be needed, for the purpose of achieving an objective$_2$ recording of the order of events, is some particular irreversible process chosen as a standard – or alternatively some quantity such as entropy which characterizes a large class of irreversible processes. The latter will be justified in greater detail in Chapter 6. For the moment it is sufficient to suppose that there is a large sample of radium whose decay has been recorded on a ciné camera over the whole temporal interval * during which time is assumed to be 'open'. The camera has also recorded various other events in its near vicinity (thus avoiding relativity effects), events such as the laying down of rock-strata, the evolution of organisms, and so on.

Consider any two momentary events, *A* and *B*, and suppose that *B* appears on a frame of the ciné film which depicts more radium as having decayed than on the frame on which *A* appears. Event *B* may then be said to be objectively$_2$ 'later than' event *A*. (By a different convention *B* could equally well be said to be 'earlier than' *A*, but the former convention is chosen to harmonize with the human sense of 'later than' without any loss of objectivity$_2$.) It would thus appear that *all* events in the vicinity of the ciné camera can be ordered in a unique sequence such that any one of them is either simultaneous with a particular state of decay of the radium, or is simultaneous with a greater or lesser state of decay. The radium sample thus appears to provide the same basis for the temporal order as was achieved in Chapter 2, but without the need for humans to make a judgment concerning 'later than' and 'simultaneous with'.

---

* Strictly speaking the radium sample should not be a closed system as otherwise physical theory would regard it as being subject to Poincaré recurrence cycles.

Clocks and calendars can also be brought into this thought-experiment since these too can be regarded as existing independently of man. Consider for simplicity a type of clock which is also a calendar – i.e. it shows the date relative to some reference date. The clock can be regarded as recording in 'the right direction' (e.g. that its hands move 'clockwise' and not 'anticlockwise') if the ciné film shows it as registering larger real numbers for greater extents of decay of the standard sample of radium. (Although it is again entirely conventional to regard increasing real numbers as being in the direction of later time.) The clock readings as they appear on the ciné film thus provide the coordinatisation of the other events which appear on the film. Furthermore, as emphasised already in § 2.4, the fact that *all* events appear capable of being accomodated within a single temporal order assures us that we need *only one* time coordinate. Time is one-dimensional.

If the setting up of time as objective$_2$ is now complete *, by use of the foregoing conceptual apparatus, it is an interesting additional issue, although one which is rather academic, whether or not the existence of irreversible processes such as radium decay says anything about 'time itself. Is time anisotropic? Does it have an arrow? Or shall we say instead that irreversibility is not a feature of 'time itself' but is merely a characteristic (and perhaps is one which is temporary during the present epoch) of certain processes, processes which occur *in* time?

The historic preoccupation of physics with reversible motions and t-invariant laws ** has tended to create a climate of opinion in favour of the second view – i.e. that the t-coordinate has no intrinsic direction. What may appear to be a similar view, but is actually very different, has been put forward by Rom Harré (1965). He defines 'time', rather as I have done, as the class of all sequences of events and as such, he says, time is neither reversible nor irreversible since it cannot itself be treated as a sequence of events. Of course I agree. But to say, as Harré does, that time is neither reversible nor irreversible is by no means the same thing as enquiring whether it is best regarded as isotropic or as anisotropic. What Harré refers to as "the class of all sequences of events" is essentially the history of the universe. If the two directions of this history are indeed distinguishable from each other, it seems best to accept that 'time itself' is anisotropic. This alternative viewpoint is well expresed by Zwart (1976) who remarks that it is important to realize that "the totality of events does not occur *in* time, but that it itself *constitutes* time." Clearly it is a question of "what one means by . . .", and it is in that respect that the point is academic. On the other hand, there is real substance to the question whether or not the two directions of the temporal order are intrinsically, as well as objectively$_2$, distinguishable from each other, and this is the issue which will be taken up much more thoroughly in Chapter 6.

---

* Whether or not it is complete will be considered further in § 5.
**This is discussed in § 6.2.

§ 5. Is the Criterion of § 4 Adequate? Let us ask about the role of the radium sample in the foregoing. On the face of it this role has been simply that of providing a very long-lived irreversible process. The radium has been used as a standard 'earlier than'/'later than' reference system in much the same way as a standard clock is used as a reference system for duration. Yet it is legitimate for this purpose only because of our confidence that it never 'reverses' itself. And how could it be said that it does not do so? Clearly only by reference to something else; as Wittgenstein (1961:6.3611) has put it: "The description of the temporal sequence of events is only possible if we support ourselves on another process."

Fortunate it is that the Second Law of Thermodynamics (in its classical and non-statistical form) *achieves a consensus* among a vast number of physical processes concerning their temporal direction. Not only radium decay, but also the mixing of gases, the equalisation of temperatures and so on, are all brought into the same picture by use of the entropy function. Thus it appears that *any one* such process can be used as a reference system, in the foregoing sense, for all the rest. But let us look at this whole matter of the consensus very carefully and consider it also in relation to the human awareness of earlier than and later than.

It was pointed out in § 2.2, that the conventional usage of words often presupposes a time direction, and that this can occur even when discourse is limited to the present tense. For instance when it is said that radium is 'decaying', that entropy is 'increasing', that the universe is 'expanding', and so on. For present purposes this difficulty can best be avoided by again using the device of the film strip and by not using any knowledge concerning which end of the strip was 'taken' at the earlier time. Thus I am now considering a film strip which shows a fair number of adiabatically isolated systems undergoing internal changes – more specifically it shows the instrument readings by use of which the entropy (relative to a standard state) of each one of those systems can be calculated for any frame.

What does a person find when he uses that strip and makes the calculations? If he considers *any one* of the entropy creating systems he finds that either the left to right or the right to left sequence along the strip corresponds to his awareness of how that process would have been seen to occur if he had observed it directly rather than as pictures on the strip; that is to say, either the one sequence or the other is a permissible sequence in relation to his own awareness of 'later than'. (Whereas if it were a question of an ideally reversible process *both* directions along the strip would have represented a permissible process.)

What he also finds – and this is the significant point – is that *all* of the entropy-creating processes show their entropy as increasing in *the same* direction, either from left to right or the reverse, along the strip. (I am leaving aside here the remote possibility of anti-Second Law phenomena – i.e. macroscopic fluctuations.*) In other words, although the various

* The extent to which fluctuations vitiate the use of the classical Second Law for the purpose of establishing 'time's direction' will be discussed in Chapter 6.

irreversible processes do not directly display their parts as being earlier or later than each other (i.e. they do not carry labels to that effect!), what they do display is a mutual parallelism of their entropy changes. It is this property of consensus which allows *any one* of the processes being used as a criterion for the temporal order, and without reducing the Second Law to a tautology.

The point is worth elaborating a little. * Let A and B be any two of the entropy creating systems shown on the film strip and let i and k refer to the numbering of any two frames (this numbering being made consecutively from one end of the strip to the other). Let S denote entropy. Then the following can be asserted quite independently of which of the two frames 'was taken' at what would be called the later moment:

$$\text{if } S_{Ai} \geqq S_{Ak}\text{, then } S_{Bi} \geqq S_{Bk}$$
$$\text{or if } S_{Ai} \leqq S_{Ak}\text{, then } S_{Bi} \leqq S_{Bk}\,.$$

Thus in either case we have

$$(S_{Ai} - S_{Ak})\,(S_{Bi} - S_{Bk}) \geqq 0 \tag{1}$$

since the brackets have the same sign whether this be positive or negative. Relation (1), which holds for all i and k and for all systems, asserts no more than the parallelism of the entropy changes over the whole duration of the processes, a point first clearly made by Schrödinger (1950).

Relation (1) is 'direction-neutral' but with it the temporal awareness of the 'observer' can now be linked up. Let this observer be *me* in the first place. What I find is that the higher entropy states are those which occur later in *my* consciousness. Thus if I had directly observed system A (and its instruments) I should have found:

$$\text{if } t_{Mi} > t_{Mk}\text{, then } S_{Ai} \geqq S_{Ak}$$
$$\text{or if } t_{Mi} < t_{Mk}\text{, then } S_{Ai} \leqq S_{Ak}\,,$$

corresponding to the i'th frame being later or earlier than the k'th respectively in my conscious awareness. Here the symbol M stands for *me* and the convention has been adopted that greater values of the time coordinate t correspond to later times in consciousness. Thus in either case

$$(S_{Ai} - S_{Ak})\,(t_{Mi} - t_{Mk}) \geqq 0 \tag{2}$$

and of course the symbol A in this relation may be replaced by the symbols B, C, etc. referring to any other isolated system. (It may be noted that the division of (2) by the positive quantity $(t_{Mi} - t_{Mk})^2$ yields $(S_{Ai} - S_{Ak})/(t_{Mi} - t_{Mk}) \geqq 0$, or in the limit $dS/dt \geqq 0$ which is the normally understood content of the Second Law.)

A further relation similar to (2) expresses the fact that any other observer is aware of the entropy states as being *in the same order* as I

---

* What follows is a paraphrase of my (1972) and I am indebted to Springer-Verlag for allowing me to use some passages from that paper verbatim.

experience them myself. Thus

$$(S_{Ai} - S_{Ak})\,(t_{Yi} - t_{Yk}) \geqq 0 \tag{3}$$

where the symbol Y stands for *you.* Yet a fourth, but non-independent, relation serves further to underline the fact that the time senses of any two observers are co-directed:

$$(t_{Mi} - t_{Mk})\,(t_{Yi} - t_{Yk}) \geqq 0\,. \tag{4}$$

Relations (1) to (4) express the complete set of entropy/time-awareness relationships within the scheme:

M	A
Y	B

where M and Y are *any* two persons and A and B are *any* two adiabatically isolated systems.

The foregoing correctly expresses, I believe, the whole of the empirical content of the kinds of experimental results on which the Second Law is based. It is the parallelism of the entropy changes, and *not* that higher entropy states occur later, which is the objective$_2$ conclusion. On the other hand the criterion concerning *which is the later event* is provided, not by the physical events themselves, but through the mediation of conscious awareness. These important points have been largely overlooked in the literature.

Notice that relation (1) would presumably remain applicable in the absence of any human observers; but the Second Law itself, to the effect that entropy tends *to increase,* could not have been formulated *in the first place* without the human awareness of 'later than'; furthermore it would not be a 'law' but for the fact, as noted above, that the time senses of all observers are indeed co-directed. Otherwise some individuals might experience entropy as increasing whilst others experience it as decreasing and there would be no publicly agreed law. To be sure good reasons can be advanced (and these will be considered in Chapter 6) why the biological or psychological sense of 'later than' is capable of being publicly agreed and why it is in the direction of an entropy increase, rather than a decrease. Yet these 'reasons' are by way of being *deductions* from the Second Law and thus it remains true, at the logical level, that the Law in its customary form is dependent on the human sense of time's direction. (see also Sklar, 1974;399 ff.)

Of course once the Law has been formulated – i.e. the decision taken that higher entropies occur in the direction of what is experienced as later – the observer is entitled to stand back from the situation and to use *any one* irreversible process as a signpost for all the rest. For example one could choose, as has been done already, a sample of radium as providing a standard earlier than/later than system – and it could be kept for reference in a Paris museum!

However it is obvious that it has been found quite unnecessary to establish any such standard. And this for the simple reason that we have complete confidence in our own judgments concerning what is 'earlier than' and what is 'later than'. (Cf. our fallible judgment concerning the metrical aspect of time where we prefer the evidence offered by a clock.) Is it not the case therefore that the criterion of 'later than' provided by consciousness possesses not only a priority, in the above sense, but also *a logical primacy* over any physical criterion such as entropy increase? This is a further facet of the whole issue and one which will now be examined.

As has just been said, we have complete confidence in our own judgments of the temporal order and this is linked to the fact that mental processes, unlike physical, display an absolute irreversibility – absolute in the sense that these processes never occur (in this case really never) in their reverse order.

Consider for example my experience of seeing a shooting star. First I have no knowledge of the event, then I see it as occurring and subsequently have knowledge that it has occurred. In a hypothetical reversal of these mental states I would first have knowledge of the event and this would precede my experience of seeing it. However the truly paradoxical character of the reversal lies not so much in this precognition (which some claim can occur) as in its supposed temporal sequel: for this would consist in the elimination of my knowledge of the event *after* its occurrence. This knowledge would be instantly and finally deleted from my mind at the very moment of the shooting star being seen!

Obviously such a sequence is entirely contrary to the facts. Once we 'have seen' or 'have known' something we can never 'unsee' or 'unknow' it. It is the complete confidence we have in this aspect of experience which explains why we regard the consciously experienced order of events *as being incorrigible* by any physical criterion.

Let us envisage the possibility that a set of events, experienced as being in the order ABC, is placed by a physical criterion such as entropy change in some quite different order such as BAC. In this case would we not find this criterion entirely unacceptable? Would it not be completely discredited, like the clock which struck the 13th hour? In other words, is it not the case that *what we really mean* by the 'later than' relation is the relation as provided by consciousness?

Quite apart from the possibility of an 'erring' physical criterion, let us suppose that we actually *saw* an event taking place in reverse – e.g. an apple 'rising' from the ground and 'attaching' itself to a tree. The time direction of these processes of 'rising' and 'attaching' are, of course, relative to our conscious sense. How then would we explain the supposed reverse event? At first it might be regarded as due to trickery, but if this proved untenable we should be driven to accept that we had suffered a visual or mental illusion. While we would thus accept the occurrence of an illusion, what I think we would *not* accept would be an alternative proposal that we had suffered a reversal of our deep lying awareness of 'earlier than' and 'later than'. This awareness we always take as being absolute and incontrovertible.

Therefore if we ask what *is* the temporal direction of events (i.e. in the sense of an identity) the answer must be that it *is* the direction of events based on the awareness of earlier and later. (This refers of course to events *as perceived* and not to events *as originated,* perhaps at a distant star.) Moreover this awareness must lie, I think, at a deeper level than the actual processes of perception or cognition, or even of the formation of memory traces. For the very fact that we can speak of 'seeing' and 'knowing', rather than of 'unseeing' and 'unknowing', shows that these processes too are judged relative to a reference. Similarly with regard to memory; its action is always appreciated as a 'taking in' of information, and not the reverse, relative to some reference process which provides our primary awareness of earlier and later. This reference process must presumably be subconscious and introspectively unobservable. For to observe it would seem to require yet a further process as a reference, and so lead on to an infinite regress.

No doubt this is speculative and in any case is unrelated to my main point. Let me conclude this section by summarising the main point: it is that the criterion of earlier and later which is offered by conscious awareness has a logical primacy over any objective$_2$ criterion offered by science. The former is "What we mean by ...". Of course in practical situations there seems never to be any conflict. Physical criteria such as entropy increase, expansion of wave fronts and so on, seem always to place events in the same order as we consciously experience them. Even so there is a useful conclusion to be drawn and this can best be illustrated by reference to Grünbaum.

Like many others he believes that 'time' is best regarded as anisotropic, which is to say that time's two directions are intrinsically distinct. But he denies that either the one direction or the other can be singled out as being *the* direction. My own view is that time is indeed anisotropic and that we can best avoid paradoxes and absurdities in its discussion by rejecting one of the two directions offered by physics and by regarding the other as *the* direction. In so far as we are sentient beings talking about our world, we should (I believe) discuss its physical changes as taking place in *the same* temporal direction as that of the 'making' (rather than of the 'unmaking') of the perceptions and cognitions which we have of this world. Such a view is, of course, tacitly accepted in most scientific discourse. If we had to think of ourselves as having the faculty of 'unobserving', as well as of 'observing', this would result in endless difficulties and confusion. We should allow therefore, within the objective$_2$ theory of time, for the one-way character of mental activity. For this activity is a natural phenomenon, like any other.

§ 6. Objective$_2$ Time and Relativity. Just how great has been the change in the concept of time due to Einstein's theory is a matter of opinion. Some authors regard the change as 'revolutionary', whilst others see it in a much less dramatic light.

In an interesting comment, Synge (1970;102) remarked that the German 'Eigenzeit', though meaning 'own time', has usually been translated into English as 'proper time', and he goes on to say: "proper time IS time in the deepest sense." And of course proper time is not affected at all by relativity! It is only when, instead of considering what is read by a clock accompanying a body (the body's *own* time), we wish to apply the notion of time to two or more bodies at a distance from each other that relativistic considerations enter in. Under such circumstances the big change made by relativity is that 'simultaneity' becomes a triadic relation. This naturally affects the measure of temporal intervals which similarly becomes dependent on the frame of reference. Nevertheless 'time' continues to be regarded as a relationship between events, and also as a continuum of instants. Furthermore the ordering relation 'later than' remains applicable – or at least as regards all causally connectible events – and relativity provides no criterion of its own for 'later than'. "What is essential", wrote Einstein, "is the fact that the sending of a signal is, in the sense of thermodynamics, an irreversible process, . . .". (Quoted by Gal-Or, 1974;437).

Let's begin by recalling Einstein's objective in developing special relativity. This was to unify Maxwell's electromagnetism with mechanics in such a way that the laws of electromagnetism, like the laws of mechanics, will be the same for all observers who are in unaccelerated motion (i.e. are in 'inertial frames of reference'), however distant they may be and however great is their relative velocity. The supposition that "the same laws of electrodynamics and optics will be valid for all frames of reference for which the equations of mechanics hold good" was the Principle of Relativity as formulated by Einstein in his 1905 paper. *

Since light is an electromagnetic phenomenon, Einstein assumed that c, the velocity of light in vacuo, has the same value in all inertial frames. A second assumption he made, for the purposes of his theory, was that no signal is available which can travel as fast as light (or other electromagnetic waves.)

These assumptions go far beyond the experimental results, even at the present day. Experiments support the view that the velocity of light, as measured at the Earth, is indeed constant, but they do not show that an observer anywhere in the universe would obtain *the same* value for c. Furthermore the measured velocity at the Earth is *the average* velocity for two-way passage – out and back again. There appear to be no means of measuring the one-way velocity of light without already assuming that widely separated clocks have somehow been synchronised. As regards the further assumption that electromagnetic waves set an upper limit to the

* References to Einstein (1905) and Minkowski (1908) are based on the English translations collected together in *The Principle of Relativity,* as published by Methuen in 1923 (and subsequently by Dover), and usually catalogued under the name of H. A. Lorentz, two of whose papers are also included.

speed of signalling, it can be demonstrated, once the theory has been formulated, that no particles can be accelerated *up to* the velocity of light. On the other hand, the theory does not preclude the possible existence of particles – "tachyons" – which *always* travel faster than light, although no convincing evidence has yet been brought forward that they do exist.

For lack of knowledge about one-way velocities of light it becomes *a matter of convention,* as will be shown below, to assume that the outward and return velocities of light are equal. Indeed in 1905 it was a daring innovation to suppose that they might be equal. For in earlier physics it had been assumed that the velocity of wave propagation is determined, either by the medium through which it passes (as is the case with sound waves), or by the source of emission, as in Ritz's theory of light. Thus, to a scientist brought up on these classical ideas, it would seem highly paradoxical that, if he were rapidly approaching a light source, the light would reach him with exactly the same velocity c as if he were stationary relative to that source, or if he were rapidly receding from it.

It's necessary to deal with this matter of conventionality before arriving at the physically more interesting issue concerning simultaneity. Reichenbach 1957) showed, more clearly than in Einstein's original paper of 1905, that a dilemma is involved in assuming *either* that clocks retain their synchrony once they have moved apart from each other, *or* that a light signal has the same velocity in both directions. Consider the first horn of the dilemma. It would be unwarranted by experiment to suppose that if clock 2 is synchronised with clock 1 at a place *A,* and if clock 2 is then taken to a place *B* and finally brought back again to *A,* it will still be synchronous with clock 1. (In fact refined experiment has shown that the clocks *don't* retain their synchrony, and that the difference of their 'times' depends on the path taken and on the speed of transport.) This being the case, there is clearly no good reason to suppose that clock 2 was still synchronous with clock 1 when it arrived at place *B.* How then can 'times' at different places be compared? The only alternative to clock transport is *signalling,* and this throws us on to the other horn of the dilemma.

According to Einstein's second assumption the fastest available signal is provided by light (or other electromagnetic propagation.) Suppose then that we use light signals to make known the reading of a clock at *B* to an observer at *A*. For the latter to calculate *how long ago* that clock reading refers to, he must know the one-way velocity of light from *B* to *A*, as well as the distance. Yet even if that distance were known, the one-way velocity could not be measured without using clocks at *A* and *B* which *are already known to be synchronised.* This takes us back to the first horn.

Thus as Reichenbach remarks: "... we are faced with a circular argument. To determine the simultaneity of distant events we need to know a velocity, and to measure a velocity we require a knowledge of the simultaneity of distant events. The occurrence of this circularity proves that simultaneity is not a matter of knowledge, but of a coordinative definition, since the logical circle shows that a knowledge of simultaneity is impossible in principle."

Nevertheless there are *limits* of times at A between which an event at a distant place B must be deemed to take place. This can be seen by supposing that a flash of light is emitted at A at a time t_{A1}, as indicated by a clock at A; this flash is reflected at B and arrives back at A at a time t_{A2}, again as indicated by the clock at A.* If the reflection event is denoted E_B, it can surely be said that E_B must be judged, in relation to the clock at A, as occurring *between* t_{A1} and t_{A2}. For if, on the contrary, E_B were supposed to occur at an instant earlier than t_{A1}, the light flash sent from A to B would have to have departed from A earlier than t_{A1} and this is contrary to fact. (There is no supposition in relativity that light can travel 'backwards' in time.) Similarly if E_B were supposed to occur at an instant later than t_{A2}, the reflected signal would have returned to A later than t_{A2}, again contrary to fact. In short, the range of uncertainty about the time of occurrence of E_B, relative to the clock at A, is the open interval between t_{A1} and t_{A2}. Let's consider the situation a little more carefully.

Now there may well be any number of events occurring *at A* between the *times* t_{A1} and t_{A2}, and these events can be locally ordered in accordance with what was said in § 4. But the *distant* event E_B cannot be said to be either earlier or later than any of them. This is because, for lack of signals faster than light, E_B itself cannot be ordered in relation to any of the events occurring at A within that interval. An order can be assigned to events at a distance from each other only if those events are *causally connectible*, and the fastest causal connection is provided by light. For this reason, the entire non-denumerable set of instants within the open interval between t_{A1} and t_{A2} is said, relative to the clock at A, to be *quasi-simultaneous* (or 'topologically simultaneous') with E_B; or alternatively E_B is said to be *indeterminate as to time order* relative to the open interval t_{A1} to t_{A2}, as measured at A. Thus in Einstein's theory there is no longer any *absolute* simultaneity of distant events, as there was in Newtonian mechanics.

As has been said, all instants between t_{A1} and t_{A2} are quasi-simultaneous with E_B – and it will be noted that this period of quasi-simultaneity is obviously greater the more distant are A and B. Einstein found it convenient nevertheless to pick out a particular instant within the open interval, for the purpose of synchronising clocks at A and B, in the special case where these clocks are not in motion relative to each other – i.e. when they are within the same inertial frame of reference. He chose the instant halfway between t_{A1} and t_{A2}. In other words E_B is assigned the time t_B, on the clock at A, as given by:

$$t_B = t_{A1} + \tfrac{1}{2}(t_{A2} - t_{A1}) = \tfrac{1}{2}(t_{A1} + t_{A2}).$$

This 'coordinative definition' clearly corresponds to the supposition that, within the given frame of reference, the outward speed of light to the reflection event E_B is equal to the return speed of light. As has been said,

* The familiar space-time diagrams are not being used in the text since a question to be raised in Chapter 4 concerns the possibility that the concept of time is falsified when time is represented as if it were a line in space.

there appear to be no means of confirming this 'one-way hypothesis' experimentally. Nevertheless a number of authors, such as Fock, seem to regard the hypothesis as being obviously true. Such authors thus represent the *anti*-conventionalist viewpoint which is in opposition to the viewpoint adopted by Reichenbach and Grünbaum (1973), as well as by Einstein himself. (The latter says clearly that the foregoing formula is intended as a "*definition* of isochronism".) As a consequence there has been much controversy about the conventionality, or otherwise, of the formula (Ellis and Bowman, 1967; Winnie, 1970), but this issue seems not to affect any of the predictions of relativity which may be compared with experiment.

Before turning to the *relativity* of simultaneity, it is worth noting that nothing in the foregoing is contrary to the possibility that the 'events' in question occur reversibly. Although it was assumed that there is a local time order at A, the terms 'earlier than' and 'later than' could everywhere have been interchanged without affecting the argument. Similarly with regard to the supposed 'emission' of light at instant t_{A1} and its supposed 'return' at instant t_{A2}; the former could equally well have been taken as a return and the latter as an emission. Thus the discussion did not presuppose any particular time order; and furthermore, although it required the temporally symmetric notion of causal connectibility, it did not require any distinction between cause and effect.

The relativity of simultaneity was explained very simply by Einstein by using the example of an imaginary train which is very long and is moving at an immensely high speed relative to the track. Two lightning flashes are supposed to strike the train, one near the front end and the other near the rear. If an observer on the ground sees them as striking simultaneously, will an observer on the train also see the flashes as being simultaneous?

Let it be supposed that the thunderbolts leave marks on the ground, as well as on the train. The mark near the front of the train is denoted F_T, and the corresponding mark on the ground is denoted F_G. Similarly R_T and R_G denote the marks made by the rear flash. Suppose too that the ground observer, G, finds by using a measuring rod that he had fortuitously been standing at exactly the midpoint, M_G, between F_G and R_G. So also in the case of the train observer; using his own measuring rod whilst the train is still moving at the same speed, he finds that he had been standing exactly at M_T, the midpoint between F_T and R_T. (See figure overpage.)

As has been said, the ground observer G *sees* the two flashes simultaneously. Since the distances F_GM_G and M_GR_G are equal, it is consistent with the foregoing convention for him to assume equal one-way velocities of light and for him thereby to *deduce* that the two flashes had occurred at the same instant, as indicated on his own clock. A corresponding deduction will *not* be made by the train observer, T, even though he is at the midpoint between the marks on the train. For T *will see* the front flash as occurring before the rear one and he will then *deduce* that the front flash *had actually occurred* before the rear one (using the same assumption as G concerning equal one-way light velocities). It is to be noticed that this is not because T would regard himself as moving towards the front flash (which is how G would think of the matter); it is rather because, at the respective moments when either of them can see the flashes, the midpoints M_T and M_G have become widely separated, due to the very high speed of the train. That this is so may be seen from the figure:

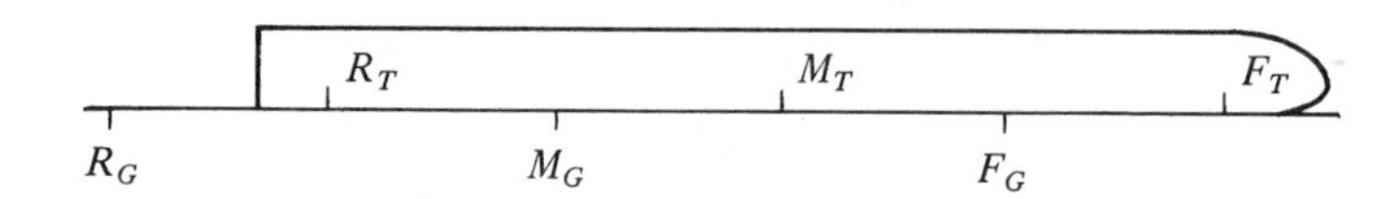

The significant point which the figure brings out is that *there is nothing relative* in the statement that the light signal from the front flash must reach M_T *before* it reaches M_G, and that the light signal from the rear flash must reach M_T *after* it reaches M_G. For this is a statement about the *order* of events in the propagation of the light, and the order of all events which are causally connectible is quite unchanged by relativity. It follows that if the ground observer finds that the flashes are synchronous, the train observer will not. And conversely.

In short, the judgment that a pair of events are, or are not, simultaneous is relative to a particular frame of reference. Simultaneity thus becomes a triadic relationship, whereas in Newtonian mechanics it was a dyadic and transitive relationship. This is the most significant consequence of Einstein's theory for the concept of 'time'.

It is not difficult to show, although this will not be done here, that there must also be a relativity of the measurement of temporal intervals. Suppose that two events both occur at the same place, P_0, and are measured by a clock at P_0 as being separated by an interval of time Δt_0. The interval Δt which will be assigned to these events by some other observer at a place P is given, according to the convention, by the expression:

$$\Delta t = \Delta t_0 / \sqrt{1 - v^2/c^2},$$

where v is the relative velocity of P and P_0, and c is the velocity of light. Since v is less than c, the denominator on the right hand side of the equation is a real number less than unity. It follows that Δt is greater than Δt_0.

Now Δt_0 is the time interval as measured at the place where the events actually occur, and this is known as the interval of *proper* (or *local*) time. This interval is evidently *the smallest* interval which can be attributed to the events; any other observer who is in relative motion (whether towards or away from P_0 is of no account since v is present in the equation as a square) will assign an interval Δt which is greater than Δt_0. This of course is the well known *time dilation* effect in special relativity *, and it

* It should be added that general relativity provides a further time dilation effect since the theory implies that a strong gravitational field can cause a clock to run slow relative to a clock in a weaker field. In the case of black holes, where the curvature of space-time due to gravitation is very great, time dilation can apparently become infinite relative to a distant clock.

can be expressed alternatively in terms of clocks appearing to run slow; observers at places such as P will regard the clock at P_0 as running slow relative to their own clocks. But of course this is a symmetrical effect, for the observer at P_0 will regard the clock at P, which is in relative motion, as running slow. It is not necessary here to deal with this apparent paradox. For present purposes it is sufficient to notice that *the only invariant time interval* between events is the interval of proper time, the interval as measured by a clock at the place where both events occur. All else is relative to a frame of reference.

Once again it should be added that, although the intervals are relative, *the order* of causally connectible events is not. What this means can be seen by considering two distant events A and B. If, at the place where B occurs, it is possible to receive, *simultaneously with the occurrence of B*, a message (by means of a light signal, or any slower signal) signifying the occurrence of the distant event A, then the events A and B are said to be causally connectible. For such pairs of events the one is quite unambiguously *later than* the other (in this instance $B > A$) not only for observers at B (or A) but also for *any* observers, whatever may be their states of motion relative to the places where A and B occur. It is only when pairs of events are considered which cannot be connected with each other by the fastest known signal that the events become indeterminate as to time order (Whitrow, 1980; 251, 325, 353).

In view of the foregoing is it correct to say, as has sometimes been said, that space and time have somehow become 'mixed up' with each other, and have lost their former distinctiveness? Now it is true, of course, that a unique $t=0$ hypersurface cannot be drawn for all observers. This has been seen already. It is also true that neither the measures of temporal intervals nor the measures of spatial intervals are invariant, except within a single reference frame. Nevertheless it remains the case that relativity does not remove the distinction between time and space. As is well known the important distinction between them which can't be obliterated is the difference of sign in the expression for the 'space-time interval':

$$ds^2 = dx^2 + dy^2 + dz^2 - c^2 dt^2.$$

In special relativity this is an invariant for all inertial frames and the difference of sign, relating to the final term, is not eliminated by substituting $-t$ for $+t$, since dt appears as a square. Closely related to this is the fact that it is the time axis, and not one of the space axes, which provides the axis of symmetry in the familiar Minkowski diagram.

To be sure, the foregoing brief account has been concerned with special relativity (S.R.), and not with general relativity (G.R.). Yet G.R. does not in any way reduce the distinctiveness of time and space (Whitrow, 1980); indeed it can be read as allowing for the existence of a unique 'cosmic time' (Whitrow, 1980;290). The situation in G.R., as well as in S.R., has been well summed up by Graves (1971;236): "... we can never convert space into time or time into space by a mere coordinate transformation; in any system that we use there will always be three space-like and one time-like coordinate at every non-singular point."

At first sight relativity seems committed to an ontology of *events,* each one of them specified within a $4D$ manifold. In his 1908 paper Minkowski wrote: "Henceforth space by itself, and time by itself, are doomed to fade away into mere shadows, and only a kind of union of the two will preserve an independent reality." In his enthusiasm he seems to have overlooked the significance of his need, on the next page of his paper, to use the concept of 'substance' in order to define his world-lines through the property of genidentity. For as soon as 'substance' or 'body', as distinct from 'event', enters into the discussion a difference between time and space becomes implicit. As Whitehead (1932;62) pointed out, if a body exists during any temporal interval (i.e. the property of genidentity) it exists equally during any part of that interval, whereas a corresponding property does not apply to space since if a body exists throughout any volume only a part of the body exists in any part of the volume. In short, the distinction between time and space is already there in the notion of a body *as existing.*

There are other differences too between time and space which persist in relativity (Bunge, 1968; Blokhintsev, 1973). Thus, as C. D. Broad put it, relativity does not break down their distinction but only their isolation.

Furthermore it cannot be said that time has become subordinate to space, or that space-time has emerged as a more significant reality than either time or space taken separately. Considering the second of these points first, it is true of course that space-time provides an invariant interval in special relativity. Yet it is also true that the notion of space-time is not usually put forward *as a primitive;* rather is it created *within* the theory on the basis of either time or space taken as the starting point. This applies as strongly to the axiomatic treatments (e.g. Latzer, 1973) as it does to the more 'physical' methods of presentation. As for the idea that time has somehow become subordinate to space, this is entirely contrary to the views of many of the relativists (e.g. Whitrow, 1980;280). It is very significant that Synge (1956, 1960), wishing to avoid introducing classical ideas only to discard them at a later stage, found it best for this purpose to define both space *and* space-time in terms of time. The single requisite measuring device in Synge's treatment is a clock (and not a clock plus a measuring rod.) Time as measured along a world-line, which is to say proper time, is taken as the basic concept of the theory, one which is already understood and to which the theory subsequently makes no alterations.

To summarise, the important function of relativity is to establish the inter-relationships which must be assigned to the spatio-temporal coordinates of systems which are at a distance from each other and which may also be in relative motion. This entails no changes whatsoever concerning events which occur at the same place P_0 as the observer. Neither the simultaneity, nor the duration, nor the order of such events is affected. The consequences of relativity in the context of this book concern what some other observer at a place P may report. In general he will not make the same judgments about simultaneity or about duration as are made by the observer at P_0. The farther apart they are, the broader is the temporal zone in which events deemed synchronous by one of the observers are not deemed synchronous by the other.

The fact that there are no means of instantaneous signalling implies that there is no 'world-wide now'. Even so, there is no ambiguity about the occurrence of events at the places where they do occur; it is simply a matter of the limitations on the transmission of the news of these events to a distance. Furthermore there is no uncertainty, as has been said already, about the *order* in which events occur so long as light signals can pass from one to another. In particular, no event which has not yet occurred for me here can be signalled to me as having already been witnessed by some other observer, whatever may be the state of his motion. As Synge (1970;89) remarks, relativity should preferably have been called 'the theory of signalling'.

In my view, relativity has therefore no bearing at all on the ontological question, to be taken up in the following chapter, whether or not there is a real 'coming into being' of things at some particular location. The occurrence of events does not become illusory simply because distant observers may not agree on their simultaneity, or on the intervals between them, for these are problems to do with signalling.

Perhaps it should be added that although 'the observer' seems to play an important part in presentations of relativity, this does not detract from the objectivity$_2$ of the theory. For in fact wherever the term is used it may be regarded as applying equally to a human observer and to a suitable recording instrument. As Grünbaum (1973;367) puts it, the physical facts which give rise to the relativity of simultaneity (as distinct from man's discovery of these facts) are "quite independent of man's presence in the cosmos and of his measuring activities."

§ 7. Conclusion. It has been seen that the main components of the concept of time used by theoretical physics (i.e. the concept of the *t*-coordinate) are as follows: (1) time is a uni-dimensional continuum of instants; (2) the ordering relation is 'later than' and its criterion is provided empirically through the existence of irreversible processes; (3) simultaneity is relative to a frame of reference and is thus a triadic relation; (4) a similar relativity applies to the duration of temporal intervals and the only invariant intervals are those which concern proper time; (5) the temporal order of all causally connectible events is nevertheless invariant.

No doubt the theory of the *t*-coordinate could not have been arrived at had there not already been available the notion of objective$_1$ time, since the latter provides the primitive terms. This is a common enough situation in science, as has been said already. For although a scientific concept may often have its origins in sensory or mental experience, this need not detract from the concept being regarded subsequently as having an objective$_2$ status. Scientific theories are abstract systems and yet, within the scope of what is usually called the 'realistic interpretation', their conceptual entities,

be these 'particles', 'waves', etc., may often be assumed to exist objectively$_2$. The theoretical entities are linked with man's observable world, the world of perception and measurement, through various *correspondence rules* (Margenau, 1950; Carnap, 1966). Let us see very briefly how these rules function in relation to 'time'.

The correspondence rule which connects 'instants' with the perceptual world resides in the relationship of instants to moments. The instant is a strictly zero temporal interval whereas the moment, in the sense in which I have used the term, is the duration of the specious present – i.e. it is the shortest interval which can be discriminated by use of the senses. Thus to each moment there is a non-denumerable set of instants and the correspondence rule is one which locates a particular instant within a set whose total duration is of the order of a tenth of a second. But of course by use of modern time-measuring devices this duration can be vastly reduced and without taking us significantly far away from what can still be regarded as 'man's observable world'.

With regard to the ordering relation the correspondence rule is that which connects the usage of a physical criterion, such as entropy change, with the human sense of 'later than'. In practice there is a 1 : 1 correspondence (and this may perhaps be accounted for by regarding the human brain as being itself an entropy creating system). So far so good. But it has been pointed out in § 5 that if we ever found a discrepancy of temporal order as between the physical criterion, on the one hand, and the human sense on the other, we would regard the former as being in error. Although this is a logical point it need not be taken as implying any real failure of the objective$_2$ theory.

Finally concerning simultaneity and the duration of temporal intervals, the appropriate correspondence rules are provided by the Lorentz transformations. Although these are highly abstract, the very successful usage of the Lorentz formulae in the study of particles moving at high velocities can leave little doubt that these particular correspondence rules achieve what they are required to achieve – i.e. the linking of abstract theory with direct observations and measurements.

So much for a brief survey. What to me is a very striking point is that there is no correspondence rule concerning 'the present'. Although the notion of 'now' or 'the present' is fundamental to the construction of objective$_1$ time, it disappears completely from the objective$_2$ theory. The issue to be examined in the following chapter is whether or not 'the present' may nevertheless be a fully objective feature of time.

Chapter 4

The Problem of 'The Present' *

§ 1. The Awareness of the Present. One of the few things that can be said with confidence about the subject matter of this chapter is that people do indeed experience a preferred moment, a 'now' or 'present'. Furthermore the sensations occurring 'at', or 'within', this preferred moment can be intersubjectively agreed; for example concerning what the clock reads or concerning your utterances and mine. Without this consensus (as we shall see later it would be a mistake to call it a 'simultaneity') it would seem impossible that we could converse with each other. In short 'the present', is quite clearly objective$_1$. The question to be raised is whether it is also objective$_2$.

A good deal concerning what it is to be aware of the present has already been described in § 2.1 and § 2.3. It was remarked that, because our experience is always *in* our present, we have no direct evidence that the past existed or that the future will exist; even our memories are present phenomena. And yet, of course, the actual content of the present, the 'now moment', is continually changing; an event which is presently regarded as being a mere possibility may 'become' a present event (and thereby enjoy the privileged status of reality for a fleeting moment), and then 'becomes' more and more past. Yet this commonplace account gives rise to some difficult questions; for instance: Is it correct to say of the past that it *existed,* or of the future that it *will exist?* And again: How shall we understand the notion of 'becoming', since (as was said in § 2.2) this appears to be spurious when applied to events?

It was also pointed out in the earlier chapter that times later than the present seem to differ importantly from times earlier than the present in regard to cognition and conation. We feel that we can know the past but not the future, whereas we can influence the future but not the past. The distinction between past and future is thus bound up in a peculiar way with the distinction between possibilities and facts; certain possibilities are capable of 'becoming' facts and they do so at each present.

---

* Much of this chapter is based on my (1978) and I am indebted to Springer-Verlag for allowing me to reproduce certain passages verbatim.

Some philosophers have been inclined to minimise the distinction between the past and the future. D. C. Williams, for instance, has remarked that the past is really as shadowy as the future, but this remark is difficult to accept. To be sure an imaginative person can visualise possible future events quite as vividly as the events he can remember from the past. Even so he does not mistake the future for the past, for the latter is characterised by a continuous sequence of memories which form a connected whole. Also we experience very different emotions towards past and future. Towards memories of the past we have feelings of regret or of pleasure, whereas towards our imagined future we have a sense of volition – i.e. the determination to influence the outcome in a desired sense.

In Chapter 2 it was also remarked that we are not normally aware of 'the present'. When I am absorbed in my affairs or when I am day-dreaming it never occurs to me to pick out a distinguished moment. Even when I look at the clock I don't usually say to myself "It is now 10.15" – I notice no more than "10.15" and this is simply to observe a coincidence of the clock hands with certain numbers on the dial.

To be aware of the present seems to require a state of awareness of being aware. As Grünbaum (1969:155–6) has very clearly put it: "M's experience of the event at time t is coupled with an awareness of the temporal coincidence of his experience of the event with a state of *knowing* that he has that experience at all. In other words, M experiences the event at t *and* knows that he is experiencing it. Thus, presentness or nowness of an event requires conceptual awareness of the presentational immediacy of the experience of the event."

Although I agree with this I believe that the present owes much of its particular character to something else – to the existence of memory. This provides the element of contrast which enables the present to be compared with what is not the present. Thus if I examine 'my present' introspectively I find that I have the awareness of the word I am now writing, but also I have memories of earlier words and of earlier events in general. If I had no memory I should presumably have a merely *punctiform* state of attention – various clock readings etc. would be experienced during various acts of attention but these would not impress me with the quality of 'being present' precisely because, in the absence of memory, I would have no power (as I put it in Chapter 2) for comparing what *is* with what *was.* All earlier experience would be wiped out at the very moment of its ceasing to be 'present' and the present would thus not stand out as being the present.

What is particularly to be noticed about memories is that *there is one which is latest* (for me now the word 'latest'). Since I don't have memories of that part of my life which is called 'future', the awareness of 'now' involves the awareness of *a terminus.*

Lest the use of the word 'terminus' may seem to be begging the issue, it is desirable to re-phrase what has just been said in terms of the minimum vocabulary proposed in Chapter 2. Within my memory there is a 'latest' one – latest in the sense that at the moment P which is my present I have a series of memories referring to moments earlier than P but none referring to moments later than *P*. This, I think, satisfactorily explicates the notion that the series of memories has a terminal member. Furthermore this discontinuity in conscious experience corresponds, as mentioned already, to the distinction between empirical facts and empirical possibilities. The former can be recorded in my memory as having the status of publicly agreed states-of-affairs; the latter are either not recorded at all in my memory, or merely as the memory of imaginings. Thus, as has been said, 'the present' or 'now' marks a division between factual states of the world and possible states of the world as observed from my own location.

Just how far down in the animal world there is any awareness of 'the present' is very uncertain. It is known, of course, that quite lowly creatures have a rudimentary memory, in the sense of being able to learn from their experiences; and this 'learning' can only be made comprehensible if the effect of the experiences is to leave some sort of *imprint* in the organism, an imprint which persists long enough to affect the creature's behaviour at the next repetition. (Agar: 1943:118 ff). Creatures other than man also have some power of anticipation; the antelope prepares for the possibility that the tiger will jump. It would thus appear that for a wide range of creatures times earlier than *P* (where *P* for man is 'a present') differ in some important respect from times later than *P*. But of course this is very far from claiming that those creatures have the sort of double-tiered awareness which is spoken of by Grünbaum. It may well be that many animals experience a sequence of moments of attention, of heightened arousal; yet the big question which has to be asked is whether man's awareness of the present is nothing more than the *self*-awareness, reinforced by the function of memory, of a similar moment of attention, or whether the present exists objectively in the world-at-large, inanimate as well as animate. And closely related to this question is the issue concerning 'becoming' – i.e. whether or not there is a real 'coming into being' of things or of events.

§ 2. The A- and B-theories. The two main viewpoints on these issues have been aptly named by R. M. Gale as the *A*- and *B*-theories respectively in view of their affinity with the *A*- and *B*-series of McTaggart. The *A*-theorists include C. D. Broad (1938), Reichenbach (1956), Whitrow (1980), Wilfrid Sellars (1962), P. T. Geach (1965), Capek (1966), A. N. Prior (1967, 1968) and Gale (1968). What they maintain, putting it very briefly, is that 'now' or 'the present' is a real feature of the world at each location, as is also 'becoming' or 'coming into being'.

The opposing *B*-theory owes much of its inspiration to Russell (1915) and has subsequently been developed by Grünbaum (1969, 1973), J. J. C. Smart (1963, 1968), A. J. Ayer (1965), D. C. Williams (1966), Goodman (1966) and Quine (1976). It has also been accepted by a number of physicists such as Costa de Beauregard (1963) and Park (1972). According to the *B*-theory, again very much in outline, events should be regarded as occurring tenselessly at certain clock times and these events are primarily 'earlier than' or 'simultaneous with' or 'later than' each other. It is these designations which the *B*-theorists regard as applying in a fully objective sense, whereas 'past', 'present' and 'future' they believe to be mind-dependent, as is also the impression that things 'become' or 'come into being'. The *B*-theory is thus essentially the assertion that the *t*-coordinate of physics provides a time-concept which is sufficient for all purposes – except perhaps for those of psychology. The *A*-theory denies this and regards the time-concept of consciousness, as described in Chapter 2, as being equally necessary for application to the physical world.

Since the *B*-theory has been more closely formulated than its rival it will be convenient, although at the risk of seeming biassed, to amplify the foregoing brief summaries by means of some quotations from the *B*-theorists – but adding, for good measure, some comments made by the *A*-theorists.

Perhaps the simplest and least ontologically commiting statement of the *B*-theory is that of Ayer (1965:170): "... events are not in themselves either past, present or future. In themselves they stand in relations of temporal precedence which do not vary with time ... What varies is only the point of reference which is taken to constitute the present, ... the point of reference, by which we orient ourselves in time, the point of reference which is implied by our use of tenses, is continuously shifted."

Gale, a vigorous exponent of the *A*-theory, has pointed out that what is misleading in Ayer (and also in Russell) is the suggestion that we are free to "orient ourselves in time" in the way we can indeed orient ourselves in space. For in fact, as Gale says (1968:202), all of our concepts of planning, intending, causing, etc. "rest on the presupposition that there is an imposed temporal schema of past, present and future that is not of our doing or choosing."

Grünbaum's version of the *B*-theory is particularly concerned with his thesis of the mind-dependence of 'now' and 'becoming'. He writes (1973:325–6): – "... the coming *into* being (or becoming) of an event is no more than the entry of its effect(s) into the immediate awareness of a sentient organism (man)." And in a later * formulation (1969:155): "... what qualifies a physical event at a time *t* as belonging to the present or as now is *not* some physical attribute of the event or some relation it sustains to other *purely physical events*. Instead what is *necessary* so to qualify the event is that at the time *t* at least one human or other *mind-possessing* organism M is conceptually aware of experiencing the event at that time."

This thesis will be considered in the following section and for the moment I will merely mention a comment made by Geach (1965). He remarks that if 'becoming' is indeed a function of mind, and of mind only, this seems to require an extreme form of Cartesian dualism. Consciousness seems to be placed in a position of peculiar isolation if, for it alone, 'the present' is real. Yet to this, and similar points, Grünbaum (1969:163) has an effective answer: "Mental events", he says, "must differ from physical

---

* N. B. Grünbaum (1973) refers to the second edition of a book originally published in 1963.

ones in some respect qua being mental, as illustrated by their not being members of the same system of spatial order. Why then should it be puzzling that on the strength of the *distinctive* nature of conceptualized awareness and self-awareness, mental events differ further from physical ones with respect to becoming, while both kinds of events sustain temporal relations of simultaneity and precedence."

Smart's formulation (1963:133) is perhaps the most metaphysically far-reaching expression of the *B*-theory. "I shall argue", he says, "for a view of the world as a four-dimensional continuum of space-time entities, such that out of relation to particular human beings or other language users there is no distinction of 'past', 'present', and 'future'." And he continues: "It is perfectly possible to think of things and processes as four-dimensional space-time entities. The instantaneous state of such a four-dimensional space-time solid will be a three-dimensional 'time slice' of the four-dimensional solid. Then instead of talking of things or processes changing or not changing we can talk of one time slice of a four-dimensional entity *being* different or not different from some other time slice."

Smart goes on to speak of the value, in his opinion, of the corresponding tenseless mode of speech in eliminating from our view of the world the anthropocentric reference which is constituted by 'the present'. As was said in Chapter 2, several other philosophers have been attracted to the *B*-theory for this reason and also because it allows of statements being made which do not change their truth value. However there are others, such as Sellars, Prior and Gale, who have expressed considerable doubts about whether tensed *A*-theory language, which they believe to be the more fundamental, can be fully translated into tenseless *B*-theory language without loss of information. "Existence statements about things", says Sellars (1962:561–5) "are as irreducibly tensed as statements about the qualitative and relational vicissitudes of things." However it would be out of place in a scientific book to go any further into these linguistic points, and neither shall I concern myself with the question whether or not there is symmetry between time and space in regard to such usages as "x is now" and "x is here". This was the subject of several interesting papers in *Mind* during 1976 and 1977, and is also dealt with in Schlesinger's book (1980).

To scientists a much more exciting feature of the *B*-theory is its evident intention to conceive the world, *sub specie aeternitatis,* as a system of spatio-temporal relationships existing between *all* events including those which, from the vantage point of now-awareness, would be said to have 'not yet happened'. As Grünbaum puts it, to say of an event that it happens or takes place or occurs is no more, according to the *B*-theory, than to say that there is a certain clock reading with which that event is simultaneous in a chosen frame of reference.

The *B*-theory, right back to Russell, has clearly been much influenced by relativity. Even so one does need to distinguish between valid conclusions drawn from a scientific theory and opinions or interpretations which are nothing more than a gloss on that theory. It is an important conclusion from relativity that the notion of a 'world-wide now' is vacuous.

So far so good. Yet it is also the case that relativity *neither supports nor disproves* the view that there may be an objective$_2$ 'now' at each location. Relativity has nothing to say on that particular issue; that is to say concerning the possibility that each genidentical body has a succession of fully objective 'nows' or 'presents' within that body's proper time, the time which is measured by a clock accompanying the body. It is significant that neither Reichenbach nor Whitrow, both of them distinguished relativists, regarded Einstein's theory as having denied the objectivity of the present at the actual *site* of an event – including the type of event which is the *receiving* of a signal at that location, even though the signal may have arrived there from a distance.

Einstein himself was equivocal on the matter – both the *A*- and *B*-theories appear to have had his support in different papers. A recent writer, Hinckfuss (1975:106), is among those who still regard relativity as having disposed of an objective present. He writes "... if the simultaneity of events is relative to a frame of reference rather than absolute, then there is no such thing as The Present." Perhaps he is right if one wishes to honour 'the present' with capitals. Yet it would be illicit to extrapolate from the non-existence of an absolute reference frame to the non-existence of 'a present' at each location. Indeed the very fact that no observer whatsoever can signal to me that he has observed an event at my own location *before* I have observed that event myself, as mentioned in the previous chapter, might seem to justify the *A*-theorist's claim that there is something absolute about the occurrence of an event here and now.

However I don't want my readers to think that I am moving forward to defend the *A*-theory. On the contrary the first point of my own I want to make is that neither the *A*-theory nor the *B*-theory, however persuasively they have been presented by their respective proponents, is properly speaking a *scientific* theory – not at least in Popper's sense. There appear to be no empirical means by which either of them might be refuted. * Indeed if one carefully studies the opposing claims they are seen to depend on *a priori* or on linguistic arguments, not scientific ones – even though the *B*-theory, as has been said, has been dubiously based on relativity.

Nevertheless there remains a lingering feeling among several authors that the issue should be capable of being resolved on scientific grounds. For example Reichenbach and Grünbaum, although opposed on the substance of the matter, both remarked that if 'now' and 'becoming' are objective this must be capable of being known to physics. Grünbaum (1969:159) wrote as follows: "... if nowness were a mind-*in*dependent property of physical events themselves, it would be very strange indeed that it could be omitted *as such* from all extant physical theories *without detriment* to their explanatory success."

Grünbaum seems to have overlooked that it is one of the important symmetry principles of science that natural laws are indifferent to a shift of

* Except perhaps if there is veridical precognition. This, I think, would be difficult to accommodate within the *A*-theory.

the temporal reference point. As Gale (1968:224) has remarked, any law or theory "is general: the statement of it is temporarily unrestricted since it quantifies over all times. Since the law or theory holds for *any* time it obviously cannot serve as a criterion for picking out some one time as *the* present: if it holds for all times it cannot hold for just one time." And therefore "... there is no physical criterion for determining what is *the* present."

In order to avoid misunderstanding it should perhaps be added that physical laws and theories are similarly incapable of picking out any unique *spatial* location; yet the existence of this analogy need not be taken to mean that there cannot be a distinguished 'present' even though there is no distinguished 'place'.

§ 3. A Dialogue. This section, and also § 4, will be concerned with developing the foregoing view that it may be impossible to bring the *A*- and *B*-theories into any real confrontation with each other in regard to empirical evidence. Although the various issues will be presented in the form of a dialogue, the reader is advised in advance that no Socrates will emerge to put things right! My objective is rather to show that no firm conclusions can be drawn about the validity of either theory.

It should first be mentioned that D. C. Williams has taken the *A*-theorists severely to task for using pictorial language, such as 'time's flow', etc. He overlooks that some of the *B*-theorists have been equally at fault in this respect. A notorious example is Weyl's notion of consciousness crawling up the world-line of its body! Even Grünbaum, who usually expresses himself so carefully, has said of events that "we conscious organisms then 'come across' them by 'entering' into their absolute future, as it were" (1973:318). I shall try to avoid all such metaphors by confining myself to the minimum vocabulary advocated in Chapter 2.

Let's allow the *B*-theorist to make the first point and this he does with an important truism: although clocks can be used to tell us 'the time' they do not also identify 'the present'. For if I were to look at the clock and report, say, "It is 10.15", my evidence for this statement is that the hour and minute hands coincide *now,* at *my* present, with such and such numbers on the dial. Thus it is the word 'is' in my utterance "It is 10.15" which contains the understanding that I am referring to a present reading of the dial, rather than to any reading of the past or to a possible reading of the future. What identifies 'the present' is not the clock, whose hands move *without discontinuity* round the dial, but rather my awareness of the moment which 'is'. For it is conscious awareness which judges that a present reading is indeed a present reading and the same applies to the reading of any other sort of instrument. Furthermore this situation cannot be circumvented by using a camera; for although a photograph would satisfactorily identify, say, a certain thermometer reading as being simultaneous with a certain clock reading it would not also contain within itself the evidence that the particular photograph is (or was) 'now being taken' – i.e. at some person's awareness of the present.

In short a statement concerning 'the time' is based on something external – a clock – but the judgment that the clock reading is *a present* reading, rather than one which appertains to past or future, is made within ourselves. Also, of course, I have no means of reading 'the time' other than at my present and this is because, as was remarked earlier, I am always *in* my present, confined to it.

The fixing of 'now' relative to clock and calendar readings, what I shall call its *dating*, thus appears to be a privileged function of conscious awareness.

What defence against this argument does the *A*-theorist offer? He may well agree that the actual dating of the present requires conscious awareness but he might go on to remark that it does not follow from this as a necessary conclusion that 'the present' is mind-dependent. It may be, he says, that 'the present' is objective$_2$ in a manner which is not concerned with its dating. He then proceeds to put forward an argument concerning the relationship of mind-states and brain-states which catches the *B*-theorists on a sensitive spot since many of them happen to be physical monists or Identity Theorists in regard to the mind-body problem.

He begins by speaking of the function of memory in regard to now-awareness, as in § 1 above, and he then remarks about our ability to make a clear distinction between memories and present perceptions, and also between present perceptions and mere anticipations. These three categories – memories, present perceptions and anticipations – are evidently three quite distinct kinds of mental states. Does it not seem reasonable, he asks, to assume that these distinctive kinds of mental states correspond, in their turn, to different sorts of physical states of the brain?

If this be accepted (and indeed it is a naturalistic way of looking at the matter) then it would appear erroneous, he says, to regard the awareness of 'the present' as being purely mental and non-physical as is claimed by the *B*-theorist. Such a view is to set up a dichotomy between mind and brain. Given the modern view that consciousness depends on a physical brain state, it would seem that there are indeed physical states of things (namely of brains) whose comparison serves to indicate 'a present'. Indeed it would appear that 'now-awareness' might perhaps be built into a suitable piece of hardware. What the hardware would require is a set of receptors, such as TV cameras, to provide the equivalent of human perceptions, a memory store which serially records an abstract of successive states of the environment, and also a predictive computer which provides anticipations of future states of the environment. Given that the hardware can also be equipped with a device for distinguishing between (a) the content of its memory store, (b) the output of its receptors, and (c) its predictions, the machine would appear to embody the necessary 'states' sufficient for it to describe the output of its receptors as being 'present'.

Taking up a different line of argument, the *A*-theorist says: Let it be assumed for the moment that the *B*-theory is right and that there is no physically distinguishable present. Thus as Smart (1968:255) put it, past, present and future "are all equally real". If this is so, asks the *A*-theorist, why does 'mind' have such a peculiarly constricted view of that which supposedly 'is'? Why is it confined to peering out at Smart's world through a sort of narrow slit, the 'specious present'? To suppose, with the *B*-theorists, that this is a true picture seems to indicate an extraordinary failure of natural selection. For surely there would be immense biological advantages in having a much broader specious present. Indeed, if it be true that all parts of time are "equally real", why are we not able to experience all-of-time-together? Why do we have the impression of a succession of brief 'nows' if physical things, including the brain, supposedly have a tenseless existence? In short, the big question the *B*-theorist has to answer is: Whence comes the awareness of a distinguished 'present' if it is not objectively real? Would it not be much more reasonable to adopt the *A*-theory and to accept that 'now-awareness' is an adaptive response to the world as it really is?

When formulating his reply, the *B*-theorist might begin by saying that he accepts much of the factual part of what the *A*-theorist has said, but not its interpretation. The *A*-theorist, looking at the matter from his own standpoint, has failed to acquaint himself sympathetically with the full resources of the *B*-theory. Let's think of some particular person (or even of the hypothetical piece of hardware) called Simon. Even though the time coordinate is undifferentiated, according to the *B*-theory, Simon will nevertheless succeed in distinguishing each successive moment of his career from the earlier moments whose contents are recorded in his memory store and from the later moments whose contents he anticipates or predicts. When each successive event occurs tenselessly, appropriate light or sound signals tenselessly impinge on Simon's receptors and each such tenselessly occurring state of perception will have the character of a particular present moment *P*. There can be no possibility of Simon experiencing 'all-of-time-together'. He cannot perceive events later than *P* because the relevant light or sound waves have not arrived. Neither can he perceive events earlier than *P*, although he can remember them since their imprints are in his memory store. Thus Simon's perceptions necessarily have a punctiform character. In short the *B*-theory does not deny that events 'occur' coincidentally with certain clock times. What it does deny is that 'the present' is anything more than a peculiarly mental registration of an occurrence. As Grünbaum had put it: "Mental events must differ from physical ones in some respect qua being mental, . . ."

As for the idea that a piece of hardware might be said to have a rudimentary 'now-awareness', the *B*-theorist does not deny this (e.g. Grünbaum 1969, Note 13). Let's look at Simon's behaviour from the outside, as a behaviourist would. If Simon says "It is now 10.15" this would be no different *as behaviour* from that which could be achieved by a speaking clock – a much simpler device than the hardware invoked by the *A*-theorist. To be sure what is difficult to understand is the double-tiered awareness, the awareness of being aware, which is possessed by humans. Yet that particular problem is concerned with consciousness in general and is not a problem to be solved within the scope of the *A*- or *B*-theories of time.

The *A*-theorist now returns to the attack from a different direction and brings out what he regards as his biggest gun. This is the simple fact that one person's 'present' is the same as another's; that is to say, the fact that different people, not in disturbed states of mind, can always agree that they are in a room together, that they are looking at the same clock and that they obtain the same reading of the clock. Indeed it is a presupposition of ordinary discourse that the communicants do, in fact, share the same present; as Gale (1968:215) remarked, it makes sense to ask a person "*Where* are you?" but not to ask him "*When* are you?".

This intersubjectivity of 'now', says the *A*-theorist, could not be explained by supposing that, out of the totality of *all* people (living and dead and not yet born, all of whom the *B*-theory must regard as equally 'real'), those who agree on the same 'now' are simply the members of the subset which consists of those who do, in fact, agree on the same now as a condition of being able to communicate with each other. For if this were true, and not just science fiction, we should expect *to see the bodies* of the members of those other sub-sets of humans whose temporal spans overlap with our own, even though their 'presents' are (supposedly) different from ours. But obviously this is not the case; *all* visible living humans in my own vicinity agree on *the same present.* Surely this provides very strong evidence, says the *A*-theorist, that 'the present' corresponds to something real in the world, that *it is impressed upon us* by a common external cause.* It is *that,* he says, which explains the intersubjectivity of 'the present', even though it cannot be excluded that it is perhaps only through consciousness that access can be gained to that particular form of external reality.

Having listened to this argument the *B*-theorist pauses for a few moments and then invites the *A*-theorist to consider two persons, Paul and Simon, (or two pieces of hardware) who are at the same location. Their receptors respond to the same external events (and necessarily in the same order for physical and causal reasons). Therefore the imprintings formed in Paul's and Simon's memory stores of any one of these external events occur virtually simultaneously. Even so, *all* of their imprintings are equally real according to the *B*-theory, since past, present and future are equally real. The *B*-theorist appreciates that he must therefore explain, on the basis of his theory, why Paul and Simon pick out, apparently fortuitously, the imprinting of *the same* external event as being present to them. He realizes, of course, that this peculiar difficulty arises only in the case of the *B*-theory since it maintains that all memory imprintings are equally real. Yet he believes it to be a difficulty which can be dismissed since it only arises on the supposition that we are asking the question why Paul's present is, in some sense, 'simultaneous with' Simon's present (as distinct from their respective imprintings). Yet that, says the *B*-theorist, is a non-question. Although we can talk about the simultaneity, or otherwise, of physical events

* See also Ferré (1970). In what follows I believe I have answered him successfully, from the standpoint of the *B*-theory.

there is no sense in talking about the simultaneity of the awareness of 'presentness', since this awareness, according to the *B*-theory, is entirely private to Paul and Simon separately. Thus there can be no sense in asking *when* Paul (or Simon) is aware of an event *E* as being present other than his awareness that *E* is indeed present.

Putting the point a little differently, each member of the Paul-sequence of perceptions can be closely paired, in regard to its content, with a member of the Simon-sequence of perceptions since these two people observe essentially the same external events. For instance Paul has a perception of the clock as reading 10.15 and so has Simon. There can be no basis for saying that these are *not* the same present, since the awareness of 'presentness', on the *B*-theory, is a private mental experience and *there is no other criterion for its occurrence.* In short if Paul and Simon are said to experience *the same* present, the word 'same' here means nothing more than that the contents of those presents transform into each other. Thus the intersubjectivity of 'now' does not disprove the *B*-theory and, by the same token, it fails to give any real support to the *A*-theory.

§ 4. Can Quantum Theory be of any Assistance? My objective in presenting the to-and-fro argument of the previous section has been to show that we are dealing with two 'theories' each of which is largely self-consistent and irrefutable on the basis of its own premises but is capable of being criticized from the standpoint of the other. Much of the argumentation of *A*- and *B*-theorists alike has been on the basis of looking at the world *as if* their own particular theory had already been established as being true.

I come now to an important contribution to the discussion put forward by Reichenbach and Whitrow which seems to offer a more directly physical argument in favour of the *A*-theory.

The point at issue between the two theories is whether 'time' *really is,* in some deep ontological sense, differentiated into past, present and future. This question might seem to be answered in the affirmative if each successive 'present' corresponds to the occurrence of some unique sort of event – some sort of event which is much more fundamental than ordinary physical changes such as doors slamming, lightning flashing and so on, since all such changes can be described equally well in *A*-theory language and in tenseless *B*-theory language. Reichenbach and Whitrow propose that there is indeed such a type of event and this is the 'becoming', or 'coming into being', of factual states-of-affairs in the physical world.

Reichenbach wrote that if 'becoming' is real the physicist must be able to know it, and he claimed that 'becoming' is in fact made manifest through the Uncertainty Principle of Heisenberg: "The concept of *becoming*", he wrote (1956:269), "acquires a meaning in physics: The present, which separates the future from the past, is the moment when that which was undetermined becomes determined, and 'becoming' means the same as 'becoming determined'".

Whitrow (1961:295) expressed a very similar view: "The past is the determined, the present is the moment of 'becoming' when events become determined, and the future is the as-yet undetermined." *

Although neither Reichenbach nor Whitrow developed their thesis at any length, the general purport of what they meant is clear: there is a basic chance element in nature, at least at the micro-level, and the moment of 'becoming', which they identify with 'the present', is marked by a transition from what is merely possible to what is factual. ** However, as will be seen below, this important attempt to provide a physical basis for the *A*-theory is by no means immune from criticism.

Consider first a couple of logical points. It has been pointed out by Sellars (see § 2.2) that it is erroneous to apply the notion of 'becoming' to events; it is *things* which may 'become' different through their events or processes of change; the events or processes themselves do not 'become'. It has also been remarked by Eva Cassirer (1972) that the issue concerning 'becoming' is logically distinct from (although obviously related to) the issue concerning 'the present'. She writes: "There is no *logical* relation (as between antecedent and consequent) between the happening of an event and the distinction between future and past. . . . it is an *interpretation* that I am putting on the word 'present' when I identify it with the 'moment of becoming' . . ." However it is precisely this interpretation which Reichenbach and Whitrow have used.

A more direct criticism is that of Grünbaum (who also quotes H. Bergmann as having preceded him). He quite rightly points out that, on the assumption of indeterminacy, the change from an undetermined to a determined state-of-affairs has *always* occurred, and therefore such a change cannot serve to distinguish what was once Plato's 'now' from what is now mine. He writes (1973:322): ". . . every 'now', be it the 'now' of Plato's birth or that of Reichenbach's, always constitutes a divide in Reichenbach's sense between its own recordable past and its unpredictable future, thereby satisfying Reichenbach's definition of the 'present'. But this fact is fatal to his avowed aim of providing a physical basis for a 'unique', transient 'now' and thus for 'becoming'."

That is surely correct. Grünbaum's criticism shows that a change from an undetermined to a determined state-of-affairs is not a sufficient condition for defining the 'present'.

Smart (1963:141) has further pointed out that indeterminism, when considered from the standpoint of the *B*-theory, means no more than that successive 'temporal slices' are not related to each other in a completely law-bound manner. This, on the *B*-theory, is just as true of what, to us, are past events as it is also true of what, to us, are future events. The real

* The same statement occurs on pg 349 of Whitrow's Second Edition (1980), but here he also accepts the cogency of some of the counter-arguments.

** It is of interest that C. S. Peirce (1935;65) preceded Reichenbach and Whitrow in holding this view. Time, he said, "has a discontinuity at the present" and this is due to the operation of chance. Stapp (1979) has put forward some very relevant ideas.

events are those which occur, each at a certain clock time, whether they are deterministically related to earlier events or not. In other words the *B*-theory is by no means committed to determinism and it allows of indeterministic events being treated as tenseless occurrences just as readily as deterministic events.

In view of these points it seems that the Reichenbach-Whitrow thesis gives little firm support to the *A*-theory. But of course this does not mean that the truth of the opposing *B*-theory is thereby established!

Are we perhaps arriving at the view that the question of the objectivity$_2$ of 'the present' is an unanswerable question? Or is it the case, maybe, that the *A*-theorist's meaning of 'the present' is synonymous, or almost synonymous, with the *B*-theorist's use of the noun 'occurrence' or of the verb 'to occur'? Notice that the *B*-theorists do not deny that events 'occur' (or 'happen' or 'take place') at certain clock times, any more than they deny that the hands of the clock are in motion. Thus it would seem that a possible gambit is to eliminate those forms of language which lead to unanswerable questions being asked. This would require the elimination of 'becoming' from the *A*-theory presentation and it would require the elimination from the *B*-theory presentation of certain metaphorical ideas such as the notion that future events are, in some sense, 'already there' and that consciousness 'comes across' them. These metaphors may well arise from the commitment of many of the *B*-theorists, to determinism – to the view, as Costa de Beauregard has put it, that 'everything is written'. For although, as has been seen, the *B*-theory does not imply determinism, the converse does not follow; indeed determinism makes such metaphors credible and to that extent weakly supports the *B*-theory.

§ 5. The Coordinatization of Time. The *B*-theory has gained much apparent support from the treatment of time in physics as an undifferentiated coordinate. In this section I shall show that this support is very weak if one (a) adopts an *epistemological* position on the issue, and (b) takes the relational theory of time seriously. But of course this does not demonstrate the *B*-theory to be false! Indeed in my final section I shall aim at achieving some degree of reconciliation between the two theories.

It appears to have been Aristotle (*Physics*, Book VI) who first represented 'time' as a straight line and subsequently the notion of time *as a coordinate*, much like the the three space coordinates, was widely adopted in classical physics. The 'times' of future events, such as eclipses of the sun, which referred to macroscopic bodies obeying deterministic differential equations, could be predicted with remarkable accuracy. The consequence of the sheer utility of the *t*-coordinate was that 'time' came increasingly to be thought of *as if it really were* something like a straight line stretching indefinitely far in either direction.

It is however entirely a gloss on the concept of the *t*-coordinate to say, with Smart, that past, present and future "are all equally real". There is nothing in physics which demonstrates this conclusion and it arises (I suggest) from nothing more than certain aesthetic or metaphysical considerations, such as the desire to attribute to 'time' the maximum of symmetry or to avoid seemingly anthropocentric viewpoints.

Let us look at the matter from an epistemological position, keeping ontological commitments to a minimum. As has been seen in Chapters 2 and 3 the *t*-coordinate is a derivative notion and is based (as is 'time' itself) on the *prior concept of the temporal order.* The significant point in the present context is that the temporal order cannot truly be said to extend 'beyond' that particular event (or set of events) which constitute the observer's 'now'. The order is a relationship between *known* events, including those events from the remote past, before man existed, whose occurrence can be inferred. Under circumstances where there are no known events there is no such relationship. It follows that the temporal order as applied to any assumed events later than now ('future time') is a schematic construction; it represents nothing more than an anticipated number of tick-tocks of the clock for the purpose of dating anticipated events; events, that is to say, which may or may not occur.

Consider this point in a little more detail. The Minkowski world-lines are often adduced in support of the *B*-theory, but let us ask in all seriousness how they can be drawn otherwise than schematically and with a flourish on the blackboard. No doubt they can be drawn with very considerable accuracy with regard to a great many sorts of 'already observed' events – i.e. events earlier than the observer's 'now'. But whenever they are extrapolated into the observer's future they are inevitably affected by a great deal of guesswork. Each such line supposedly describes the space-time behaviour of a genidentical body or particle. To attempt to draw the line beyond any actually observed events is to assume that the particular entity's future behaviour is reliably and accurately predictable. But in general this is not the case for it cannot be known that the body or particle will not undergo some event which is utterly unforeseeable. If it is a star it may suffer an explosion; if it is a nuclear particle it may undergo a disintegration and thus give rise to two or more divergent world-lines. Since the 'times' of such events cannot be known the future world-lines cannot be drawn.

In short there is an important epistemological distinction between the recorded or inferred events of the past, on the one hand, and the uncertain events of the future, on the other. This is a distinction which has been largely glossed over by some of the *B*-theorists. Furthermore, because the future events are indeed uncertain, there is no relational basis for the construction of a future space-time 'grid' – there is nothing on to which it could be pegged, so to speak. The mapping against known events is missing, and this is what I meant about taking the relational view of time seriously.

It follows that the assumed continuum of the temporal order (its assumed decomposition into 'instants') differs from, say, the continuum of the real numbers in one very important respect. One real number is related to another by being 'greater than', and similarly one instant is related to another by being 'later than'. Yet this latter relationship is vacuous except by reference to concrete events, and thus to *empirical states of the world* (including states of consciousness). For if there were no difference anywhere in the universe between the physical and mental states corresponding to supposedly different times, there would be no basis for believing that they were not one and the same time. The next second or minute has therefore a different logical status from, say, the next digit of Π. Even though this digit may not have been worked out, it has a definite value in a way in which the physical content of 'future time', as far as human knowability is concerned, does not. For clearly there is no *logical* necessity that all change in the universe, including the ongoing of clocks, will not suddenly cease. The extrapolation of the temporal order (as distinct from the t-coordinate) beyond the present is thus defeated by the logical failure of induction.

What has been said is in no way contrary to the continuum theory of time since Cantorean concepts are just as applicable to a series which is bounded at one or both ends as to a series which is endless. And neither is it contrary to relativity. To be sure as soon as one postulates the existence (in some sense) of 'time', as distinct from the local temporal order, one is adopting a universe-wide concept and this has to be accommodated to the facts of signalling. The t-coordinate must then be assimilated to the space coordinates, as is done in relativity theory. But this does not affect my point which is that the t-coordinate is a schematic construction, invaluable to science though it is. To endow it with overmuch reality is an instance of the false reification of 'time'.

Notice that it would be impermissible, from the epistemological point of view which is being adopted, to pose the question: "How fast does the present move on?". Such a question would be to assume that 'time' is indeed correctly regarded as a coordinate, one which already exists (so to say) ahead of the present. So too if it were said that the present 'advances' into time. There is no 'existing ahead' and no 'advancing' according to the viewpoint I am adopting, which was also that of Broad (1927).

But of course this doesn't preclude speaking of the time and date of some anticipated event such as an eclipse of the sun. For what one would be saying is simply this: that if the sun and moon continue to move along their present paths and if clocks too continue to behave deterministically, then an eclipse is likely to be observed at such and such a reading on the clocks. This understanding of the t-coordinate as being simply a convenient tool for purposes of prediction does not carry with it as a *necessary* consequence that past, present and future must be equally real. Such a belief, if one wishes to adopt it, requires some additional ontological commitment. From the bare epistemological standpoint, the present is the moment P such that there are no moments later than P to which any physical content can be ascribed; there is nothing 'real' later than P and in particular one should not think of future time as 'existing'.

C. F. von Weizsacker (1973) expressed a similar view when he wrote: "One would expect a measurable time not to be a real-number parameter but a counting operator, perhaps counting the facts, that is the events which have happened up till now."

With this idea in mind let us think of any moment j during Plato's life. For him that moment was a divide – a divide which separated his factual knowledge of earlier events from his imagined possibilities of later events. But of course that moment j is not a divide for me, it is not one of my 'nows'. For me Caesar's crossing of the Rubicon is a fact, but it could not have been so for Plato. For Plato, and for those who lived contemporaneously, the class of real events was the class E_j of all *known* events – events that is to say whose occurrence was either earlier than or simultaneous with the moment j. For me and for my contemporaries a much larger class is real; larger because the class E_j which was accessible to Plato at the moment j is now (at my present) a sub-class within my own (or at least potentially so assuming complete recording). Thus each successive 'present' corresponds to a larger class of events which are real. *

In thus suggesting that the temporal order should strictly be regarded as having a terminus at the present, this I regard as what can be justified on the basis of what is knowable. Yet it is perhaps desirable to go beyond this with some sort of ontological claim. The reason arises not so much from our confidence that future events will indeed occur as from our knowledge, scientific knowledge, about the past – the remote past.

Suppose one considers the past at a period before there were any humans capable of having the awareness of 'the present'. All our scientific knowledge of that period, as obtained particularly from geology, indicates the reality of a temporal order pertaining to that period. For instance, one says that some particular kind of rock was laid down 'later than' some other kind, and so on. Indeed we have means of *dating* the events of that period.

The 'terminus' I have spoken about is, as has been said, a terminus of knowledge relating to the future – it is an epistemological terminus. If, on the other hand, one considers the remote past before there were humans it seems that one must *either* (a) retain the epistemological viewpoint and thereby declare that the notion of a terminus, a 'present', is simply devoid of meaning in relation to the temporal order then pertaining, *or* (b) make an ontological claim in the form of asserting either that there was, or that there was not, a progressively changing 'present' throughout that period. With this I come back, once again, to the basic question this chapter is concerned with. If I have made any advance at all within this section it has, I think, been nothing more than the high-lighting of the distinction between the epistemological and the ontological issues.

§ 6. A Naturalistic Viewpoint. Towards the end of § 4 it was suggested that, when the B-theorist says that an event 'occurs', this may be almost the same as what the A-theorist means when he says that an event

---

* Russell (1940;101), in an interesting passage, remarks that this notion of the class-inclusion of recorded states can be used for the purpose of achieving a serial ordering.

'is present', or 'was present' or 'will be present'. This may be a useful clue to the attainment of some degree of reconciliation between the two theories. Even so I am inclined to go along with the *B*-theorist in thinking that 'present' can only be used in the sense of being present *to* a living creature. In other words the notion of 'the present' should not be attributed to the temporal order in the absence of sentient beings.

Our awareness of time differs from our awareness of space in this important respect: in *any one* state of awareness we are conscious of things or events as being at different places but not as being at different times. This is because the state of awareness *is itself an event* and, as such, it has the characteristics of any other sort of event. In particular it has a position within the temporal order and can be assigned a clock time t. A person cannot have a direct awareness of external events other than those which occur *at* his present, and thus (as was said already in § 3) perception has a punctiform character. In short, any external event, if it is to be perceived, has to coincide, in a temporal sense, with a mental state which is also an event.

Notice too that 'the self' is temporally extended, but is not experienced as being spatially extended. The self is a coherent succession of mental states, or events, rather than a collection of objects at different places. This is not to say, of course, that the self doesn't have a spatial location; it does; yet this location is that unique reference frame which is *me* (or my body). My own mental events are normally * experienced as being concentrated at the point of origin of this reference frame, and therefore appear as having no spatial extension.

It is reasonable to suppose that other living creatures also have mental events, even though they are probably lacking in the double-tiered awareness referred to by Grünbaum – i.e. the awareness of being aware. Thus a naturalistic way of looking at the whole problem is to consider the continuity of organisms downwards from man to simpler forms. The having of a 'present', when seen in this light, is surely an outcome of natural selection. It does not *necessarily* signify any 'becoming', or 'coming into being', of things, but could be explained, in *B*-theory terms, as the occurring of some appropriate mental state coincidentally with the occurring of changes in the environment.

Indeed it has been established by stimulus-response experiments that other creatures as well as man are subject to that shortest discriminable time duration which in man is known as the 'specious present'. Some quoted figures for animals lie in the range 1/30 to 1/4 second (M. Grene, 1974;282, from Adolph Portman). ** Continuity arguments suggest that the same applies to still lower forms of life even if these display only a relatively slow response to environmental changes. Think for instance of the slow opening of the petals of a crocus. The phototropic mechanism requires fairly prolonged states of affairs, such as would be described by the

* I say 'normally' because of the occasional occurrence of 'out of body' experiences.

statement "the sun is shining", rather than event-like happenings such as would be expressed by "the sun has just come out". Yet this is probably nothing more than a matter of degree – and indeed if the process were displayed by time-lapse photography the plant's response would be seen as being in every way comparable to the animal's.

Thus the most straightforward view of 'the present' is that which was already mooted in § 1 – i.e. it is the mentalistic correlate of a state of greatly increased physiological and psychological arousal, a state of heightened intensity of perception which is requisite to the living creature preparing for action in the face of external occurrences. Such states are, of course, detectable by physiological methods and are objective$_2$ states of living bodies. Presumably they exist on all planets where there are sufficiently advanced forms of life.

Whether or not man's own *self*-awareness of the present is also to be regarded as objective$_2$ now appears as a somewhat semantic issue. In § 2.1, objectivity$_2$ was defined as that which can be held to exist or to occur quite independently of man's thoughts or emotions, *or* of his presence in the world *except* where, in the latter case, it concerns his body. All scientists would agree that man's bodily events, such as his heart beats, are fully objective; it would be consistent with this to maintain that his brain events, including his awareness of 'the present', are also objective$_2$. Even so, it will be clear from the foregoing naturalistic account, that there could be no *awareness* of the present if either man did not exist or if there were no comparable creatures anywhere in the universe having man's faculty of conceptual awareness.

In short I would say that although some of the exponents of the *B*-theory draw too sharp a distinction between 'mind' and 'non-mind', the theory correctly exphasizes the significance of the consciousness of self. On the other hand the awareness of a thing or of an event is not a necessary condition for that thing or event's reality; for obviously there are many 'realities' of which we are not yet aware but which we shall certainly regard as being real, and as *having been* real, as soon as they are discovered. Thus the foregoing considerations still leave quite unanswered the question whether or not 'the present' has some sort of reality quite apart from the awareness of it.

** It would be of great interest for these researches to be extended since there are two conflicting expectations as to whether the 'higher' organisms might have a greater or shorter duration of the specious present. On the one hand, if they are able to focus their attention more sharply than the lower organisms, the duration of their specious present might be expected to be shorter; on the other hand, if the duration is determined by information processing (cf. Wheeler, 1967;234), the higher organisms might be expected to be capable of a longer specious present. (See also § 8.2.)
Perhaps it should be added that, although I have described the specious present as the shortest discriminable time duration, there are several alternative definitions. Indeed the duration of the specious present is dependent on the particular technique which an experimental psychologist uses in measuring it.

Since everything so far seems to imply that this is an unanswerable question it is desirable to attempt seeing the issue in an entirely different perspective.

If any general conclusion can be drawn from the preceding chapters I think it is this: that the time concept cannot be separated from a consideration of consciousness without its becoming impoverished. Furthermore this need not be thought of as a subjectivist conclusion so long as consciousness, *pace* the behaviourists, is regarded as a fully objective feature of the natural order.

As was stressed in § 1.1, 'time' should not be conceived as some sort of *existent* since it could hardly be said that time exists in time. Thus it is not a physical entity having its own intrinsic properties – i.e. determinate characteristics which are available to be discovered. It is rather an abstract relational concept. As such its *attributes,* and also the number of them, depend on the level of the particular phenomena where specifically temporal relations need to be established. It is a question of different levels of description.

Following this line of thought, it would be reasonable to suppose that the time-concept, as we pass from the kinds of phenomena which are the concern of physics to the kinds of phenomena which are the concern of biology and psychology, must be endowed with *a progressively increasing degree of richness.* At the level of atomic phenomena, where all processes (with the possible exception of those relating to the weak interactions) appear to be t-invariant, a sufficient time-concept is the directionless coordinate which is used in theoretical physics. But as soon as we are concerned with largish aggregations of particles and with macroscopic phenomena it becomes necessary to take account of seemingly irreversible changes – even if these are regarded, on a vastly extended time scale, as if they were temporary fluctuations. Thus 'history' enters in and a richer time-concept has to be used, one in which there is now an intrinsic distinction between time's two directions.* Even so neither the direction of increasing entropy nor the direction of decreasing entropy can yet be singled out as being *the* direction. As has been seen in § 3.5, it is only at the level of conscious awareness that a unique direction makes its appearance. And it is also at this level that the time-concept becomes still further enriched through the manifestation to conscious awareness of a *unique moment* – the now or present. This provides the distinction between past and future and therewith the distinction between factual states-of-affairs and possible states-of-affairs.

Although I have spoken of enrichment this could lead to a misunderstanding of what I mean. In my view it would be erroneous to regard the more comprehensive time-concept of conscious awareness as if it were obtained by the adding on of further features to a supposedly more basic concept originating in physics. The reverse, I believe, is closer to the truth;

---

* This transition between micro- and macro-concepts of time has been particularly clearly brought out by Landsberg (1972). It will be discussed in greater detail in Chapter 6.

the t-coordinate is obtained by a process of discarding – by an elimination of those particular features of time-as-it-is-to-conscious-experience which are not required for the limited purposes of physical theory. In other words there are aspects of the more comprehensive time-concept which need not be made use of in the theories of micro-events and of reversible motions.

It would be a prejudice arising from the great success of physics in its own proper field if the pared-down or minimal notion of time which it uses were to be regarded as 'more true' in some sense. In the case of a relational concept it is not a question of truth but rather of adequacy and significance.

The value of regarding time in this way – i.e. as a multi-level relational concept – will be brought out more strongly in Part II. My ideas on the subject have been greatly stimulated by the interdisciplinary studies of Fraser (1975), although the foregoing differs somewhat from his own presentation in points of detail. Prigogine too has reached similar conclusions. In his Nobel Lecture (1978) he remarks: "... the development of the theory permits us to distinguish various levels of time: time as associated with classical or quantum dynamics, time associated with irreversibility ..., and time associated with 'history' ... I believe that this diversification of the concept of time permits a better integration of theoretical physics and chemistry with disciplines dealing with other aspects of nature."

Part II

Temporal Processes

Chapter 5

The Interplay of Chance and Causality

§ 1. **Introduction.** The following chapters will be about three distinctive kinds of temporal change: processes occurring in inanimate nature; processes of the living; and finally processes related to consciousness. Questions to be asked will include how these processes are to be characterised, and to what extent they are really distinct from each other. First, however, I need to take up a very general issue: Whether these kinds of process are fully 'law-bound', or whether they exhibit chance effects or randomness. This is the long-debated issue about determinism versus indeterminism.

In my view there is something unreal, and indeed almost ridiculous, about both alternatives! Take determinism. This is to say, as C. S. Peirce (1935;30) puts it, that the smallest curlicues of the letters I am now writing have been completely determined since the world first began. What, if anything, can this mean? Does it mean that God ordained my curlicues, or that they were 'written' into the initial state of the universe, so to say, if there was an initial state? And in any case how could it possibly be known that these things are true? Determinism is a doctrine which smacks of medieval theology with its demand for certainty and absoluteness in all things.

Yet indeterminism too is a peculiar doctrine. It seems to imply, to put the matter in an anthropomorphic fashion, that an atom or electron literally 'doesn't know what to do'. And again indeterminism may seem to be a defeatist, or even an irrational, position – one which shakes the scientist's confidence that he can achieve a progressively greater degree of understanding of natural processes (Nagel, 1961 b; Wilson, H. 1961).

We seem furthermore to be up against the law of the excluded middle; for if determinism is a proposition either it, or its negation, must be true. But perhaps the matter isn't as simple as that. It could be that determinism and its negation are thoroughly unsatisfactory concepts and will eventually disappear from philosophical discourse; the supposed proposition has no application, like talking about unicorns. Or again determinism and its negation should perhaps be regarded as categories of thought, and not of nature, categories which will be superseded as the result of better understanding. Indeed this is the trend in much of modern science where 'laws' are regarded as probabilistic, and not as implying certainty.

Even so, it is not easy to break away from an established philosophic tradition and find an entirely new vocabulary embodying novel concepts. There are words such as 'necessity' and 'chance', belonging to the doctrines of determinism and indeterminism respectively, which are firmly entrenched in ordinary language and cannot easily be dispensed with.

For this reason determinism and indeterminism have to be dealt with in their own terms, and my aim is to show how the issues can be clarified by considering them *in a temporal context.* My thesis, in brief, is that the only reasonable form of determinism is the predictive (and not the ontological) kind, and is the claim that an event E becomes more and more accurately predictable the shorter is the temporal interval between the moment of prediction and the moment t_E at which the event occurs. During the same interval the chance, or random, factors may correspondingly diminish in importance. In other words predictive determinism refers to a sort of limiting or temporally asymptotic claim. Yet in some instances even this relatively modest claim goes too far, for the asymptote may be missing. That a scintillation due to a sample of radium will appear on a screen at exactly 10.00 hours is no more predictable at only 1 second before the hour than it was at 10 seconds. In such instances it can only be said that the event is 'determined' when it actually occurs, and thus becomes an observed fact.

It is the ontological form of determinism, the doctrine that events are completely determined in advance, which is mainly questioned in this chapter. Historically such a view had a strong theological underpinning, but from the time of Galileo and Newton obtained its strongest support from the existence of laws of nature. Yet this is dubious support if only because ontological determinism – or just 'determinism' for short – makes an absolute claim whereas the 'laws of nature' are tentative and provisional; they express partial states of knowledge at a particular period. As such the laws are continually changing; Newton's laws gave way to Einstein's, and so on. Furthermore the laws are admittedly approximate; no modern scientist would regard $pV=nRT$, or indeed any other gas 'law', as necessitating the *precise* behaviour of any actual gas. To base determinism on supposed 'laws' requires the conjecture that *there really are* (that there *exist,* in some sense) hidden scientific laws of unlimited precision. To make this ontological jump about something so clearly the product of man's intellect as our actual scientific laws appears quite unwarrantable.

And again if it were supposed that the curlicues of what I am now writing have been determined, in accordance with these putative hidden laws, ever since the world's beginning, this is to require that these hidden laws are changeless in themselves. This is a further ontological claim and is not in harmony with the current state of scientific belief. For our actual manmade scientific laws, quite apart from the revision and improvement of them due to continued research, may apply only to particular epochs; in fact it is widely supposed that the 'laws' of physics which appertained to the first small fraction of a second after the big-bang may have been quite different from those which have appertained to its subsequent development.

My own view is that scientific laws should be regarded as mental things * – continually improving in accuracy and in predictive power though they are – and have little bearing on the merits of determinism when this is considered as an absolute doctrine. As Cassirer (1956;213) put it: "... the original fallacy of the entire causal problem consists of considering laws themselves as a kind of reality, ..."

As for real events, these happen as they do happen. It is fortunate for us that natural events, including our own actions, display a very considerable degree of regularity, but I can see no kind of *necessity* in this. It is simply a matter of how the world is, and of how our minds are attuned to it, that a good deal of regularity can be discovered. But it did not have to be so. The world might have been such that there was a much looser 'fit' between successive states-of-affairs. In other words there might have been much greater randomness, and correspondingly much less scope for the discovery of lawlike behaviour.

Although the word 'chance' will be used extensively in what follows, let me say that I use it in an adjectival sense, and not substantively. In my view the notion of chance is descriptive of the manner in which a pair of successive states-of-affairs may be related to each other. Thus to assert that 'chance' is real is not to say that a something called chance exists; it is simply to say that there are pairs of occurrences which do not stand to each other in a relationship of necessity. And of course this doesn't mean the absence of *any* relationship between those occurrences; it means a relationship of probability, rather than of certainty, and is thus not contrary to causality when this is understood in a broad enough sense (Born, 1949;102). In short, 'chance' is relational concept, just as much as is necessity (Kneale, 1949;115).

It is part of the general thesis of this book that some degree of chance, understood in this relational sense, is as necessary to an adequate understanding of the world as is a considerable degree of regularity. Without regularity we could not achieve any real comprehension of natural phenomena; but without some degree of chance, however small, there could be no appearance within the world of anything which is genuinely novel. It would be to say, with Aquinas, that everything is "already present in its causes", and although this idea may have a certain attraction to the theoretical physicist, *qua* physicist, it would be profoundly at odds with the evolutionary sciences, as well as with psychology and ethics.

§ 2. 'Causes'. A few words need to be said about 'cause' and 'effect' even though these notions are little used in science. As is very familiar the

---

* Scientific laws have to be capable of supporting counterfactual conditionals, and these, says Rescher (1973), exist in the mind only.

term 'cause' in everyday language refers to some abnormal event or to some intentional act which, together with the remaining circumstances, appears to give rise to the 'effect'. In other words attention is fixed on whatever aspect of the circumstances functions as a trigger and this depends on a process of recognition. No necessity is involved here; although a spark is recognised as a cause of ignition, it does not necessarily result in that since the other circumstances may be unfavourable. To be sure one can adopt a more comprehensive notion of 'cause' as including the prevailing conditions as well as the ostensible releasing event, but the consequence of this move is to change the discussion into a discussion of determinism. For whereas the terms 'cause' and 'effect' refer to events, determinism refers to a supposed relationship of necessity between total states-of-affairs.

In some examples (e.g. the depression 'caused' in a cushion by the action of a weight) cause and effect appear to be simultaneous, but in general some part of the cause always precedes the effect. The weight has to be put on the cushion in the first place, as otherwise there is no effect. The distinguishing between cause and effect thus presupposes that time indeed has an 'arrow', whereas the more general notion of causal connectibility, as used in § 3.3 and § 3.6, does not require this presupposition and neither does the thesis of determinism. These refer to temporally symmetric relations between *states*.

The 'arrow' which is presupposed in causal statements appears always to be consistent with the arrow provided by thermodynamics since all 'real life' changes contain an element of irreversibility, however small. For instance a spark which 'causes' an ignition initiates a spontaneous entropy-creating process and similarly when a weight sinks into a cushion its loss of potential energy appears as heat. Thus causality does not provide an independent criterion of time's direction. This will be discussed further in § 6.4.

Since the notion of 'cause' is of little utility in science it is surprising that it continues to be discussed so extensively by philosophers (e.g. by Mackie 1974; Von Wright 1974). In my view rather little that is scientifically interesting has come out of these discussions. The notion of 'cause' is ambiguous precisely because, as has been said, it fastens on only one feature of the total circumstances of an event. Indeed it can be used in a sense which is actually contrary to the thesis of determinism. As Eddington put it: "How can I cause an event in the absolute future, if the future was predetermined before I was born?"

A number of authors (notably Harré 1970; Harré and Madden 1975) have made the metaphysical assumption that there are 'generative mechanisms' and 'natural powers'. No doubt there is something to be said for this. Take for instance the pressing of a lever; as soon as the one end has been pushed down it seems to follow, by a sort of physical necessity, that the other end must rise. The rigidity of the rod, together with negligible friction at the fulcrum, supplies a generative mechanism concerning why the lever must rotate. Similarly regarding 'natural powers'. Rain has the power to wet the ground and nitric acid has the power to dissolve copper. And of course they will do so provided all relevant conditions are satisfied; the ground must not be too hot and the acid must not be so cold as to be solid. But I am not convinced that the notion of generative mechanisms and of natural powers is as useful as these authors seem to suggest. Admittedly things have properties (and thereby have an efficacy for causation); yet it is the entire set of property-values of all entities involved in an interaction which defines a state-of-affairs and which thus supposedly 'determines' what happens. Moreover 'powerful particulars' (to use Harré and Madden's term) have their 'powers' partly by virtue of what *we* can make them do; for obviously a given object, say a stick, can be a causal agent in many different ways. Therefore to speak of the efficacy of specific agents for particular events seems to me less scientifically profitable than to speak of the ordinary physical *properties* of those agents – properties which man exploits in a variety of ways according to his faculty of imagination.

One further point about 'causes' before I turn to determinism; the assumption that there is a cause, or that there are a plurality of causes, of an event does not exclude the possibility that there are also factors which may properly be called chance factors. Consider the example of a man who is killed, when he is walking along a street, by a tile falling off a roof. It would be entirely consistent to maintain, on the one hand, that the cause of the fatality was the breaking of a nail which held the tile; and also to maintain, on the other hand, that it was a matter of chance that the man arrived at that particular location at that particular moment. Similar considerations apply in many laboratory contexts. There is no contradiction in saying that, although the causes of the instability of atomic nuclei is understood, the individual atomic events within a disintegration process form a random series.

§ 3. Predictive Determinism. The traditional notion of determinism contains both epistemological and ontological elements and this is well illustrated by a definition due to Russell. Determinism, he says, is "a belief that what happens in the world dealt with by physics happens according to laws such that, if we knew the whole state of the physical world during a finite time, however short, we could theoretically infer its state at any earlier or later time."

In this Laplacian-style definition there is a conjunction of three distinct features: (1) the assumed law-boundedness of the world; (2) the knowability of its state; (3) the possibility of using this lawboundedness and knowledge for the purposes of prediction and retrodiction. (See also Feigl and Meehl, 1974).

Considerable clarification of the concept has, I think, been achieved through the making of a distinction between the epistemological and ontological elements. This distinction ws drawn by O'Connor (1956), Bunge (1959;1967), H. Wilson (1961), Bradley (1962) and Watkins (1974), although Cassirer (1956;63) reports that it was already known to Helmholtz. Bunge puts the matter thus: "... since predictability is a human ability and not an objective trait of nature, it is necessary to distinguish *ontological determinacy* (=determinacy *in re*) from *epistemic determinacy* (=complete knowability). *

Bradley makes the point that the truth of the ontological thesis is a necessary condition for the truth of the epistemic thesis. Although this is logically sound it is also the case that scientific predictions are made with ever increasing accuracy, as scientific knowledge advances, *without* the scientist being necessarily committed to a belief in ontological determinism. The situation in quantum mechanics is a case in point: the properties of ensembles can now be predicted in circumstances where the older theories were useless, but at the same time the confidence in determinacy *in re* has been severely shaken!

Furthermore the history of science from Galileo up to the end of the 19th century indicates that the continued belief in ontological determinism was not so much a precondition of scientific discovery as a *consequence* of science's predictive power. To be sure Galileo and Newton approached their studies with necessitarian convictions arising from theology. Yet Laplace's stupendous claim arose out of the immense success of Newtonian mechanics and he would perhaps not have made his claim if he could have known about quantum theory.

The predictive power of science can therefore be discussed quite separately from determinacy *in re* and that is what will be done in the present section.

Now it is true, of course, that scientific prediction is often very imperfect and there are many familiar reasons why this is so. There may be insufficient knowledge of all relevant phenomena, but even when the relevant phenomena are known the available 'laws' may be only approximate. Then again there may be a lack of the complete data which are required, in a concrete instance, for the purpose of applying the laws. This situation often prevails in regard to the predictions of events taking place under non-isolated conditions. Take for instance the predicting of the time and date of the falling of a particular leaf from a particular tree; apart from requiring a great deal of data about the leaf itself, this would also require knowledge about future states of the weather and the boundary conditions for the prediction of these states would necessitate a truly fantastic amount of information. Non-isolated phenomena are affected, in principle at least, by *all* events within a portion of the past light cone. If at time t_1 an event is to be predicted as occurring at the later time t_2 this would require a knowledge of all parts of the past light cone which could have an influence on that event up to the instant t_2. However, because of the finite speed of transmission of the fastest signals, not all of those

* In Bradley these are referred to as 'causal determinism' and 'predictive determinism' respectively. Watkins uses the corresponding terms 'metaphysical determinism' and 'scientific determinism'. In this chapter I use 'ontological determinism' and 'predictive determinism' as conveying most clearly what is meant.

influences could be known at t_1. Furthermore an event which is 'caused' by a high energy photon coming from outer space (such as may occur in a genetic mutation) could not be predicted at all. The news of the approach of that photon could not travel faster than the photon itself, and thus leave no time for the making of the prediction (Watanabe 1972).

In spite of these difficulties classical physics held out the hope that a convergence of predictions * towards an ultimately perfectly precise prediction is possible in principle, or at least in situations where the effect of influences coming from distant parts of the universe is negligible. A familiar example is the treatment of the solar system as an isolated gravitational system. The achievements of celestial mechanics, using this kind of assumption, led Reichenbach to remark that classical determinism had its origins in the successes of astronomy.

Recent developments in science have, of course, greatly reduced confidence in the possibility of ultimate convergence. No doubt the most familiar of these 'anti-deterministic' developments is the emergence of quantum mechanics. The significance of this need not be discussed since it has been the subject of a vast literature. It is sufficient to remark that 'uncertainty' at the atomic level is readily compatible with an *apparent* determinism at the macro-level for two well-known reasons: (1) the effects with which Heisenberg's Principle is concerned become less and less significant the greater the mass of the particles in question; (2) where there are large assemblies of atoms and molecules, as in all macroscopic systems, the consequence of the 'law of large numbers' is the appearance of highly reliable statistical laws, even though the individual micro-events may be affected by Heisenberg uncertainty. The emergence of lawboundedness as the result of statistical averaging was, of course, well-known in advance of the advent of quantum mechanics, for example in the kinetic theory of gases. As Borel (1928:290) has pointed out, an apparent determinism at the macro-level does not entail determinism at the micro-level, or vice versa.**

However I would not wish to advance any argument against predictability or against determinism on the strength of quantum mechanics, if only because this is a *theory* and some other sort of theory may be successfully developed. What is surely a much more significant recent development (in the present context) is displayed by certain *empirical* features of nuclear physics. Experimental research has disclosed an ever-increasing number of 'particles' or 'resonances' and there is no present indication of any finality. It may well be that there are layers below layers below layers – and perhaps without end; the 'qualitative infinity of nature' as Bohm (1957) has aptly called it.

In short there is no present prospect that the scientist will ever attain knowledge of a truly ultimate level such that information about its

* The notion of the convergence of predictions is very fully described by Reichenbach (1956).

** An instance of the 'vice versa' would be the intersection of two independent causal chains, each of them supposedly micro-determined.

properties would allow of all nuclear phenomena becoming predictable. But if so, the convergence of prediction cannot be taken for granted in connection with *any other kinds of phenomena.* For these cannot be confidently separated off, as it were, from what happens in the nucleus. To be sure the gross motions of the stars and planets do happen to be capable of being separated off to an apparently high degree of accuracy; but there are other sorts of macroscopic phenomena which are unpredictable to the degree that nuclear phenomena are also unpredictable.

A good example concerns the cooling rate of the stars and planets which until a few decades ago was assumed to be the same sort of cooling as for a lump of hot metal. With the discovery that the sun is fuelled by the conversion of hydrogen into helium and that heat is generated within the earth by radioactivity, all previous estimates of cooling rate had to be drastically revised. The significant point is that we cannot know that we now have final knowledge on this matter. For that would be to suppose, *inter alia,* that we have the finally complete information about the smallest particles or resonances and about all of the energy-yielding reactions which can occur between them.

Quantum and nuclear phenomena are by no means the only areas of science where there are no present indications that there can be a convergence of predictions towards a perfectly precise limit. Another such area is biology and indeed the whole field of the life sciences.

One of the reasons is that no two organisms are exactly alike; each organism is an individual and differs in detail from its fellows, even in the instance of 'identical' twins. For obviously it could never be said of John and Jack that their bodies contain exactly the same number of atoms bound together in precisely the same molecules. However similar may be their genetic make-up, it would be entirely fortuitous if 'nurture' had resulted in every detail of their bodies being the same. In the case of any pair of cells the shapes and sizes of the internal organelles are far from identical. Darwin himself remarked "No one supposes that all the individuals of the same species are cast in the same actual mould." Furthermore this factor concerning individuality seems to become more marked the higher the level in the evolutionary scale. A natural consequence is that biological 'laws' are inevitably statistical and thus do not apply with certainty to each individual. Nor could physico-chemical laws be used in their place; for the information required for their detailed application could only be obtained through a dismembering of the living organism. In short neither type of law can be used to predict the behaviour of an individual organism except as a probability.

Another important reason why biological phenomena are very imperfectly predictable arises from the immense sensitivity of organisms to quite minute environmental changes. As is well known the eye can respond to a very small number of quanta * and similarly just a few molecules of

* The evidence is described by Pirenne (1952). Professor R. A. Weale in a private communication tells me that it is now believed that a regular periodic signal of *only one* quantum per unit summation time of the rods or cones is sufficient to act as a detectable stimulus.

certain substances are sufficient to stimulate the sense of smell. Yet, once the threshold stimulus occurs, the nervous system of any sufficiently developed organism is capable of *amplifying* it enormously and thus of giving rise to a macroscopic response. This means that a stimulus which is so minute as to be unpredictable in regard to its occurrence (being dependent on fluctuation phenomena or perhaps even on quantum uncertainty) is nevertheless capable of leading to an act of behaviour.

The figure summarizes, albeit in a schematic and qualitative manner, what has been said about the inaccuracy of prediction. If one follows the catenary curve from left to right, the inaccuracy first diminishes due to the diminishing significance of Heisenberg's Principle, together with the effect of the 'law of large numbers', but it rises again on the right hand side due to the factors which have been mentioned as being operative in organisms. The trough of the curve forms what might be called 'The Newtonian Region'; indeed the purpose of my diagram is to emphasise that Newtonian mechanics was applied to systems which allowed of a very high accuracy of prediction.

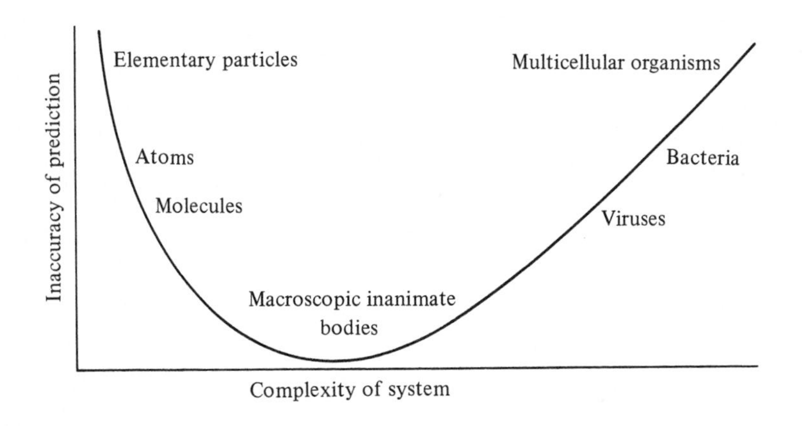

It was perhaps fortunate for early modern science that Galileo and Newton concentrated their pioneering work on the motion of inanimate macroscopic bodies. The almost perfect predictability of these motions produced a psychological climate which provided scientists with a great impetus towards further advance. And yet it was also unfortunate in so far as it created an expectation that this advance would take place by use of the same methods as had served Newton so well – i.e. by the setting up of deterministic differential equations. "The thesis of determinism", wrote A. J. Ayer, "has lived very largely on the credit of classical mechanics."

The effect of the immense success of Newtonian theory had thus been to reinforce the desire for absolutism which had been inherited from theology and mathematics. As Cassirer (1956:77) put it: ". . . every natural law had been regarded as possessing the essential property of an indwelling necessity, a necessity that excluded every exception." Burtt (1932) and

Buchdahl (1969) have described the history of these necessitarian ideas in science.

What might have been the alternative to the use of deterministic differential equations? Patrick Suppes (1974:281, 295) remarked that one of the surprising things in the history of science was the very late development of the theory of randomness and probability, and he attributed this to the factors which have been mentioned about the belief that truth could not be otherwise than absolute. There was thus a long delay in science before it was realised, with the success of Clerk Maxwell's kinetic theory of gases, and the work of Boltzmann and Gibbs in thermodynamics, that there can be perfectly good statistical laws which apply with great precision to aggregates, even when the behaviour of the individuals which form the aggregates may not be predictable. Correspondingly there was also a long delay before it was understood that an important distinction needs to be made between the determinacy of a theory (such as Newtonian mechanics) and the determinacy, or otherwise, of actual events or states-of-affairs in the world at large.

Of course prediction, whether by use of deterministic theories or by use of statistics, remains an essential objective in science. But it is always limited by what information we have, and *can have.* As was said in one of my earlier books (1975): "... is it not apparent that an *entirely complete* description of any material object can never be achieved? An exhaustive description of a conceptual entity, such as a triangle or an ellipse, is always possible, but the same does not apply to objects belonging to the real world; something would always be left unsaid." It is the gaining of ever more knowledge and understanding, limitless though the process is likely to be, which (subjectively at least) is one of the sources of the view that 'time is creative'.

§ 4. Ontological Determinism. Turning now to the ontological proposition, determinacy *in re,* it has been well argued by many authors that this is not a falsifiable hypothesis. Since there is an abundant literature, just a brief outline of the reasoning will be sufficient.

There are two main candidates for the actual statement of ontological determinism:

(a) If a state A of any sufficiently isolated system is followed by a state B, the same state A will always be followed by state B; *
(b) Every physical event is entirely governed by categorical (i.e. nonstatistical) and temporally invariant laws.

---

* Concerning (a) as an instance of the Principle of Sufficient Reason, see Bunge (1959;229), Franklin (1968) and Berovsky (1971). No doubt Leibniz' Principle is no more falsifiable than is (a) itself; and further the 'sense' of the Principle has been criticised by Sklar (1974).

The main question which arises from (a) is how the particular system can be known to be 'sufficiently isolated' – unaffected, that is to say, by changing external factors, such as might be due to the entry of penetrating particles, or to the influence of external fields. For if, on some particular occasion, the scientist were to observe that state *B* did *not* follow state *A* he would normally assume that there were just such factors causing a departure from nature's normal regularity. In other words he would assume the truth of (a) in order to show that his system had not, in fact, been fully isolated. This *reverse usage* of (a) precludes putting it to the test! But of course this is not to dispute the immense heuristic value of *assuming* (a) to be true. For it was just this assumption which led to the discovery of previously unknown phenomena whose reality was subsequently verified by independent methods. A particularly striking instance was the discovery of Neptune; certain variations in the orbits of the already known planets indicated that the solar system, as then known, did not constitute a fully isolated gravitational system. This use of (a) for purposes of deduction was then triumphantly vindicated by direct sighting of the new planet. But of course successful applications don't prove a general principle to be universally true.

There are other familiar reasons why (a) is unfalsifiable and these have been discussed by Frank (1949) and Nagel (1961 a:317). But let's turn briefly to (b) and see if this fares any better. Now in a certain sense (b) can be made trivially true, as shown by Russell, by constructing a mathematical function, albeit of a fantastic degree of complexity, which is capable of correlating *all* known events up to a given instant. But of course such a function is not what a scientist means by a 'law of nature' and neither can it be depended upon to predict successfully beyond the given instant (Scriven 1957; Berovsky 1971).

Leaving Russell's function aside, it can readily be seen that (b) is just as much a will-o'-the-wisp as is (a) as regards refutability. For suppose there is an apparent counter-example – some phenomenon such as ESP which is not yet understandable on the basis of existing theories and 'laws'. It could always be said that this is because the requisite theories or laws have not yet been brought to light – or, in other sorts of examples, because the data required for applying the theory or laws have not yet been obtained with sufficient precision. Thus once again, as with (a), there is a circular situation – one in which an apparently non-lawlike phenomenon, together with the assumed truth of (b), is used as evidence that new or better laws, or better data, remain to be discovered. From one point of view the unknown is inviked in order to 'save' (b); from another, (b) gives valuable guidance towards scientific discovery!

It will be noted that ontological determinism is related to the hypostatization of scientific law and is also related to the idea that the material world is essentially passive and has no initiating powers of its own; 'causes' are never originated but must be capable of being traced back to an inscrutable First Cause, or in an infinite regress. More will be said about this important aspect of the matter in the following section.

If in the light of the foregoing it were now said that determinism, or the unrestricted Principle of Causality *, is nothing more than an invaluable methodological principle, "an injunction to continue to seek causes", the objection might be raised that, if it is indeed so valuable, the likelihood is that it asserts something truthful about the way-the-world-is.

Against this position two things can be said. The first is that determinism – or rather causality – may in fact express a partial truth about the world without expressing the whole truth. We should not expect that our ontological assertions, formulated as they are in human language with all its limitations, are capable of being in any respect more absolute in character than our empirically tested scientific conclusions. Hanson (1958) has put the point very pungently: "Causes certainly are connected with effects, but this is because our theories connect them, not because the world is held together by cosmic glue."

The second point is that quantum theory, according to many of its expositors, provides a ground for an ontological thesis which is the *contradictory* of determinism. Quantum theory, it is said, supports the view that there are at least *some* events which are not determined in the sense of the propositions (a) or (b); at the level of atomic events the Copenhagen school maintains that there is a basic indeterminacy and we only obtain the impression that there are deterministic laws from the fact that all scientific laws are really statistical in character and represent the aggregate behaviour of vast numbers of micro-events each of which occurs only in a probabilistic sense.

Be this as it may, what I hope to show in the following section is that, as soon as the matter is considered in a temporal context, determinism will be seen to be altogether excessive in its claims as compared to its contradictory.

§ 5. The Time Factor. Let's think in the first place about how the time factor affects *predictive* determinism, taking as our example the case of the man killed by a falling tile.

Consider the situation 10 seconds before the event; the man is then walking towards the location of the possible impact and the nail holding the tile has perhaps not yet broken. There may still be many circumstances whose effect would be to falsify a prediction that he will be killed; for instance he may slip on the pavement or he may stop to look at a shooting star. All such circumstances would need to be precisely known for the purposes of a hypothetical prediction at 10 seconds and clearly it would

* 'Determinism', in my usage of the term, is synonymous with *universal* causality – i.e. with what is also known as The Principle of Causality (Schlick: 1961-2). But causality, *tout court,* I regard as having a broader meaning, one which allows for the possibility of understanding the causes of an event, and/or of achieving a statistical relationship, without implying that each individual micro-event is separately determined. Max Born (1949) has a similar usage but Bunge (1959) regards causality as a sub-class of determinism, the former applying only when one event may be said to *produce* another.

require a fantastic amount of data. On the side of the man's history it would require a knowledge of all factors, such as shooting stars, which might cause him to stop, or to slow down, and, on the side of the tile's history, it would require the data needed for predicting that the nail will indeed break during a particular very short interval of time, together with the data required for showing that frictional forces will not prevent the tile from sliding off the roof.

Consider now the situation only 1/10th second before impact; already the tile is falling through the air and the man is taking his final stride. A greatly reduced amount of data, in particular that which specifies the locations and velocities of the approaching bodies, may now be sufficient to allow of a fairly reliable prediction that impact will occur. Even so the additional prediction that the man will be killed, rather than merely injured, remains very uncertain. For this will depend on the character of the impact, on the man's recuperative powers after injury, and so on.

This example shows the significance of the time factor in prediction – and it may be noted that it is not at all inconsistent to adopt causal talk for the purpose of demonstrating the difficulties. As was said in § 1, the shorter is the temporal interval over which a prediction is projected the greater in general is the reliability of that prediction. Even so an event, in any non-isolated system, becomes certain only when it does in fact occur. For up to the very instant of that occurrence there may still be a virtual infinity of circumstances which preclude reliable prediction by means of a finite prediction process (Popper, 1950).

The same point can be made by use of a second example: the classical calculation of the trajectories of gas molecules, in a container, from a supposedly known initial state. Of course computational difficulties would be immense, but there is also a difficulty of principle which was expressed in a dramatic manner by Borel (1928;174). He asked himself: What would be the effect on the gravitational field at the earth's surface if a mass of one gram was displaced through a distance of one centimetre on a star at the distance of Sirius? He showed that the change would be no more than by a factor of 10^{-100}, but nevertheless this would be sufficient largely to falsify the predicted trajectories of the molecules in the container on the earth's surface after a period as short as 10^{-6} second! Thus, in certain sorts of situations, there is extreme difficulty in achieving sufficient freedom from minute external influences to ensure reliable prediction even over exceedingly short temporal intervals.

This is not to say, of course, that there are not other sorts of situations, for instance those relating to the motions of the stars and planets, where predictions over long periods of time are very accurate indeed. There are good reasons why this is so, for celestial bodies are extremely massive and are therefore insensitive to chance variations in the influences arriving at their surfaces. Furthermore such influences are greatly attenuated over the vast distances which are involved. As Bunge (1959:100) has remarked, science would be impossible if everything in the universe had to be related to everthing else.

Distinction must evidently be made between insensitive kinds of phenomena, on the one hand, and highly sensitive kinds on the other. Catastrophe theory deals with the minute changes which are often sufficient to tip the balance between a process developing in two or more alternative ways. Very delicate effects of this kind are known to occur in meteorology – for instance in the initiation of cyclones. So also in biology: a high energy photon arriving from outer space is able to initiate a genetic change which subsequently becomes amplified in the form of a whole population of slightly altered organisms.

So much concerning the time factor in prediction and I turn now to the same factor as it concerns ontological determinism, i.e. the claim that events really are fixed unalterably, in some mysterious sense. This thesis, *qua* being ontological, must clearly apply to circumstances *in general* – that is to say whether or not these circumstances can be specified by means of a small number of variables of state, and whether or not the system in question is isolated. Furthermore as regards the temporal factor it would be erroneous to think of ontological determinism as being true only in a limiting sense – i.e. as becoming 'more and more nearly true' as the temporal interval diminishes. We are concerned here with an *absolute* doctrine and for this reason the temporal factor is very different from what it is in the instance of predictive determinism. It is a case neither of *approaching* truth as a temporal interval diminishes, nor of *being* true for some particular temporal interval, arbitrarily chosen. The thesis must be assumed true for *all of time,* or not at all.

Consider the example of the man killed by the falling tile. If ontological determinism is accepted at its face value, it cannot be said that the man's death suddenly became determined at, say, a second before its occurrence. If the event was so determined the total circumstances must have been just as much determined a year in advance as at a second in advance; and they must have been just as much determined at 10^n years in advance as at one year in advance; and so on right back to the beginning of time, if there was a beginning. Every detail about the man and his birth, and every detail about the nail and about all other circumstances, must have been as if already 'written' during a possible infinity of time.

Such a notion raises some quite extraordinary problems. For if there was 'a beginning' was it in some sense a well-defined beginning such as would allow all subsequent states of the universe being determined by it? The sort of 'beginning' which is pictured by the big-bang theorists is hardly encouraging to such a view. Or alternatively, if there was no beginning, does it make sense to suppose that everything has been determined for an *infinite* time, without reference to any 'initial' state since it would be quite arbitrary to choose one? *

Perhaps it might be claimed that the foregoing objections are deceptive since it is not a question of an event being *directly* determined by the

* Compare Kant's Third Antinomy. In his discussion Kant assumes, of course, that the 'laws of nature' are absolute.

circumstances 10^n years earlier; it is rather a question of a continuous sequence of states-of-affairs such that each state determines another at an infinitesimal time later. Nevertheless the objection still remains; for what determinists believe is that infinitesimal changes can be integrated, in the sense of the calculus, over possibly infinite durations of time.

Consider an analogy. A rigid steel rod of, say, 1 cm diameter is almost unbendable if it is only 10 cm long. But suppose it is 10^n miles long? Very minute impacts would cause it to wave about like a blade of grass. The actual constitution of matter is not such that there is complete rigidity over indefinite lengths; rigidity is an idealization. Similarly, I suggest, the 'determination' of one state of affairs by another is also an idealization and is only to be taken as a good approximation to the extent that, in any actual instance, prediction over finite time intervals is found in fact to be a good approximation.

The matter can be looked at from a different angle. Although much emphasis is rightly given by scientists to lawlike behaviour, the ontological determinists go much further by regarding the laws as *necessitating* events to occur as they do occur. Yet they offer no evidence or reasons why the laws themselves supposedly remain unchanged. And indeed the possibility that the 'laws' do change has been given serious attention by Dirac, Wheeler and others in the context of cosmology. If this possibility is accepted, how would the thesis of determinism cope with the situation?

Consider the set L_1 of all known laws of nature. If these are liable to change it might perhaps be claimed that there is another set of laws, L_2, at a deeper level which are capable of describing how the set L_1 changes. Yet the set L_2 may also change and if so the claim would need to be extended by postulating a further set of laws, L_3, at a yet deeper level. This leads therefore to an infinite regress and the significant point is that nothing can be said by the determinist against this possibility.

The extraordinary magnitude of the claim put forward by ontological determinism will now be clear. It assumes that the occurrence of any and every event has been determined, in a continuous sequence, throughout the entire period of a possible infinity of time. This supposition is peculiar enough in itself, but it becomes the more so if the thesis also requires an infinite regress of laws.

By contrast the claim put forward by ontological *in*determinism is a very modest one. For this, as the negation of ontological determinism, is nothing more than the thesis that there may be at least one event which is not necessitated by any earlier (or later) state of affairs. Let's proceed to consider the reasonableness of the view that there are some events of this kind.

§ 6. 'Chance'. Some seven different usages of the word 'chance' have been distinguished by Ayer (1969;110), but most of these need not concern us here. For present purposes it is sufficient to distinguish between weak and strong senses of the term. In the weak sense a chance event is simply an event which cannot be predicted in a particular context, but may be so predicted in a different context, where more knowledge is available. In the strong sense a chance event is an occurrence such that there is no scientific backing for the view that it could *ever* be predicted, other than as a probability.

As noted in § 1, the notion of chance is to be understood adjectivally – i.e. as a characteristic of a relationship. Furthermore chance is relational in a two-fold manner: first in being concerned with the manner in which a pair of events, or more correctly a pair of total states, are related to each other; secondly in being relative to states-of-knowledge. For instance, as Brown (1957) has put it, what appeared to me to be a chance event that Jones visited me today, was not a chance event to him.

There are at least two important groups of phenomena which display chance in the strong sense: (1) Fluctuation phenomena (such as Brownian motion), particle decay, absorption and emission of radiation by atoms, 'activation' in chemical reactions and many other similar processes; (2) Changes and events within non-isolated systems arising from the coming together of two or more independent 'histories'. Examples from both of these groups will now be discussed.

(1) *Particle decay.* The following argument will be familiar to most of my readers but is worth describing since it makes very clear how weak is the deterministic alternative to the supposition that particle decay involves 'strong chance'.

It is known from experiment that the decay occurs according to the Rutherford-Soddy 'law':

$$N = N_0 \exp(-\lambda t) \tag{1}$$

$$\text{or} \quad dN/N = -\lambda \, dt \tag{2}$$

where N_0 is the number of undecayed particles in the sample at any chosen initial instant $t=0$, N is the number at a later instant t and λ is a time-independent constant. The latter is very simply related to the 'half-life', $t_{1/2}$, the time for half of the original number of atoms to decay: $t_{1/2} = \ln 2/\lambda$.

Equation (2) tells us that the fraction, dN/N, of particles which decay during a period dt is proportional to the duration of this period but *is independent of the age of the atoms.* This is a very striking empirical result and it can be re-expressed in the language of probability theory (using the frequency interpretation) by saying that the probability of a particle decaying during a period of, say, one second remains the same throughout the whole of its actual life. Although this was originally established for the case of radio-active decay, it also applies to the decay of more elementary particles such as neutrons, mesons, etc. (Omnés 1971; Burcham 1973).

Before considering how the Rutherford-Soddy law may be accounted for, it should be mentioned that there is a further empirical result which has been very firmly established. This is that the value of λ (or of $t_{1/2}$) for a particular radioactive element is quite unaffected by changes in the external physical conditions. In any radium sample (of adequate size) just half the radium atoms which are present at any instant will decay after a further period of 1600 years and this value of $t_{1/2}$ is not altered in the slightest by subjecting the sample to very high temperatures or pressures, or by changes in the state of chemical bonding of the radium atoms or of their closeness of proximity to each other. The half-life remains the same whether the radium is present as the pure metal, as one of its salts, or as a very dilute solution of one of these salts. This independence of external conditions seems to show that the factors which give rise to decay must lie within the individual particles. Yet these factors are evidently not affected by the age of the particles.

The matter may be put even more strikingly. Consider any chosen initial instant: the probability that any particle in the sample will decay during the next second is just the same as the probability that it will decay, if it has not already done so, during the same period of a second at a time a thousand, or even a million, years later.

It may be added that this probabilistic picture is supported by the fact that the decay rate is subject to fluctuations. The number N of undecayed atoms in a sample is normally many millions or billions. But, in samples where N is relatively small, fluctuations in the value of dN/dt can be detected – and indeed were predicted to occur by Von Schweidler quite early in the history of radioactivity. λ must therefore be regarded as the *most probable,* rather than as the absolutely constant, value of the fractional decay constant for the particular radioactive element. Von Schweidler had made the successful prediction that certain features of the process of decay are unpredictable!

At present there appears to be no prospect whatsoever of a theory being set up which would allow of the precise instant of decay of a particle being predicted. Indeed such a theory would appear to require a knowledge of the precise 'state' of the particle at some initial instant $t=0$, and it is hardly conceivable that such knowledge could be obtained because of the limits set by quantum theory. But of course this is not to say that decay is causeless in the sense of being inexplicable. There can be good theoretical reasons why certain particles are unstable which yet do not allow of a prediction of the time at which an individual decay event will occur.

Let us ask nevertheless what assumptions an ontological determinist would need to make towards dealing with the situation. Since external factors are irrelevant to the decay process, he would need to assume that, within each sample of particles, there is one particle pre-determined to decay at, say, precisely 10^8 seconds after some chosen initial instant, that there is another particle pre-determined to decay at 10^8+1 seconds, and so on for all particles in the sample. And of course he would also need to assume that each one of these individual 'timings' is exactly correlated with the others in such a manner as to ensure that the right proportion of particles have the appropriate lives in order to obtain agreement with the observed decay law, equations (1) and (2). This is to assume a Leibnizian pre-established harmony at the initial instant!

The difficulties for the determinist become even worse compounded if he is asked how he chooses that initial instant. Take the case of a radium sample. The radium is extracted from uranium ores and a sample consists of atoms which were formed from their radioactive precursors. The various atoms in the sample will have been formed at widely separated instants. The determinist must therefore bring the precursors into his picture and, when this is done, the causal chains in which he believes cannot be said to have any 'beginning' at a particular instant. Indeed it would be contrary to the determinist's whole position to suppose that 'causes' can ever be initiated *de novo.* The pre-established harmony must therefore be taken back to the beginning of the universe, if there was a beginning, and if not must be taken back an infinite time.

The deterministic account therefore requires very elaborate and peculiar hypotheses; in particular the assumption that the particles have been 'timed' over a possibly infinite duration in such a way as to exactly mimic what would be expected on the basis of the alternative assumption that the individual decay events have a chance relationship to each other. In my view it is a much more economical hypothesis to suppose that they do indeed have such a relationship.

Bradley (1962) has argued in an apparently contrary manner as follows: "... if atoms are the kind of thing which are capable of decaying at some instant or other, then it is logically necessary that any statement asserting this or that specific disintegration time for a particular atom should be either true or false – and determinately so. That is, it is logically necessary that if an atom decays at some time or other it decays at a definite time."

Of course I accept this, but it is really beside the point. For one reason because the hypothesis that decay is random requires no assertions about *particular* atoms. One cannot carry out experiments on particular atoms, labelled as it were. For another reason because Bradley's remark amounts to little more than the truism: "What will be, will be." What he is concerned with is the *logical* necessity that at any instant a particle has either decayed or has not decayed. This is not to be confused with the thesis of ontological determinism which, if it can be decided at all, can only be decided on the basis of empirical evidence concerning the-way-the-world-is.

So much about particle decay as an example of 'strong chance'. The same conclusion may be drawn from many other events and processes: fluctuation phenomena, emission and absorption of radiation, 'activation' in chemical reactions, and so on. Consider the latter very briefly in order to show its similarity to the example of particle decay. The successful existing theory of chemical reactions assumes a random distribution of the lives of activated molecules. An alternative deterministic theory would require the assumption of a pre-established harmony of lives and would also require that *each and every time* a sample of the reacting substance is prepared by the chemist the pre-established harmony is always such as will imitate randomness. This is surely most unlikely since the individual occasions of preparing the samples must differ from each other very considerably in matters of detail.

The assumption that chance events occur in isolated systems is clearly in no way to deny the statistical law-boundedness of phenomena; on the contrary it often provides the simplest basis on which the law-boundedness *of ensembles* of entities can be accounted for.

(2) *The convergence of 'histories'.* In non-isolated systems 'strong chance' can also be manifested as the coming together of two or more unrelated causal chains. One good example which has been quoted already is that of the man who is hit by a falling tile; for here the event depends on the whole past history of the tile, and of the nail which holds it, and the whole past history of the man. Yet there are countless other examples. A mutation induced by a cosmic ray particle is dependent on the history of the organism as well as that of the source of the particle; for the event does not occur unless the organism and the particle happen to come together at the same place at the same moment.

Or take the simple example of the fall of a penny. It may well be the case that if one knew the precise force with which the penny is projected, together with precise details about air currents, etc., the fall of the penny could be predicted. Yet that would be an instance of asymptotic predictive determinism as discussed earlier. Ontological determinism, on the other hand, assumes that these *conditions* of tossing were determined, and were so for the whole of time. Rather than making this far-reaching assumption concerning the various intersecting causal chains it seems much more reasonable to assume randomness. As Gillies (1973;134) puts it: "It may be true that given the exact initial conditions of a particular toss its result is thereby determined, but the exact initial conditions of tossing will *vary randomly* from one toss to the next."

Criticisms of determinism on similar lines have been made by Landé (1961) and Watkins (1974). In general it may be said that the assumption that apparently quite independent histories are not really independent but were co-determined at the world's beginning seems utterly fanciful. Indeed it would be difficult to reconcile such an assumption with big-bang theory, quite apart from quantum mechanics. The big-bang theorists do not postulate a 'beginning state' which was perfectly orderly, like a crystal at the absolute zero. On the contrary the assumed initial state was at an immense temperature and seems already to have contained considerable entropy; furthermore it has been concluded that the existence of a particle horizon ensured that arbitrarily small regions of the early universe were causally isolated from each other and were thus uncorrelated (Davies 1974;112, 179).

To be sure this is speculative cosmology. Yet the significant point is that the determinist would have to assume, far more speculatively, that the motions of every particle and every photon were not in fact disorderly but were as if they contained precise instructions, sufficient to 'determine' every subsequent collision process between particles and photons during the whole period of some 10^{10} years between the big-bang and the present epoch. It is the need for this kind of assumption which makes ontological determinism incredible.

My own view, as has been said already, is that the only reasonable form of determinism is the predictive kind and this does not make an absolute claim. It is a matter of degree. By this I mean that we are usually presented with *mixed situations* such that, for a particular system *S*, there are a number of factors internal to *S* which allow, at the least, of probabilistic predictions concerning events within *S*, together with other factors external to *S* whose effects may be utterly unforeseeable.* Thus I would fully agree with how the matter has been expressed by Lamprecht (1967;79): causality and contingency, he says, are not alternative principles between which philosophy has to choose, but are "mutually consistent and supplementary principles which together describe fairly the nature of events."

§ 7. The Amplification of Micro-Events. For the purposes of later chapters it is useful to conclude with a few words about how chance events at the atomic level become amplified, often by a sort of cascade effect.

One commonly distinguishes between 'events' and 'processes' – the former as referring to sudden changes which appear to imply a discontinuity (e.g. in the first time derivative) and the latter as referring to smooth changes, as when a chemical reaction takes place in a beaker. If so, the question arises how temporal discontinuities are to be accounted for.

Notice first of all that 'events', in the sense which has been mentioned, seem to fall into two groups: (a) Those which are clearly due to the amplification of single atomic occurrences, as for example in the operation of Geiger counters, cloud and bubble chambers and photographic emulsions. (In the latter for instance it appears that the formation of each grain of silver, containing millions of atoms, is triggered by a single quantum); (b) Macroscopic events such as the breaking of a nail, the rupture of a dielectric or the sudden sliding of a rock on a mountain side. It is the thesis of the present section that many, if not all, members of the second group are just as capable of being traced back to the occurrence of single atomic events as are the members of the first group.

In general macroscopically discontinuous change arises from *the breakdown of a metastable state.* The nail breaks when it is under stress and when its material contains a number of minute cracks (as is normal); and it breaks at a particular moment when a critical concentration of stress develops, within a region of atomic dimensions at the head of one of those pre-existing cracks, perhaps as the result of a spontaneous fluctuation. Similarly a dielectric medium, such as air, will rupture and a spark will occur when there is a state of strain which exceeds what Maxwell called the electric strength of the medium. Particularly well-understood instances occur in the breakdown of metastable phases such as supersaturated

* For an interesting treatment of ontological *in*determinism as a matter of degree see Von Wright (1974;13 ff).

solutions or supercooled melts. Formation of the new and more stable phase takes place when an appropriate nucleus makes its appearance and this can be due to 'seeding' from outside or to the spontaneous formation of a nucleus exceeding a critical radius due to a random fluctuation. *

Consider *any one* occasion of this sort and let us ask whether or not breakdown has occurred within a given interval of time Δt. The answer must be either yes or no. For it is not here a question of the probabilities within an ensemble of systems or occasions, but rather of the behaviour of *a single example* of a metastable system on a particular occasion. Within the interval Δt either the critical condition has been reached or it has not. And the further logic of such situations, I want to suggest, is that a single occurrence *at the atomic level* – e.g. a fluctuation in the position of a single atom, the breaking of a single chemical bond, or the gain of a single extra quantum of energy – is able to tip the balance between the non-occurrence and the occurrence of the critical condition.

For let it be supposed, on the contrary, that one were to assert that as many as, say, 100 atoms or quanta are required to tip the balance on the particular occasion. This would be to say that 50 are insufficient and that 90 are insufficient, and further that 99 are insufficient. But if 99 are insufficient whilst 100 are sufficient, just one has tipped the balance! **

The hypothesis which I am thus advancing is that many instances of macro-events are triggered at the level of individual micro-events and are therefore subject to the random character of thermodynamic fluctuations or of quantum phenomena. A similar view has been put forward by Gillies (1973;137) who suggests that all macro-randomness is an amplification of randomness at the atomic level.

In connection with this notion of the amplification of chance occurrences it may be remarked at this stage that biological systems, as has been said already, are endowed with singularly powerful amplifying mechanisms. What is also remarkable about organisms is that they have very effective means of *selection*; the chance events which come their way are selected and only a fraction are chosen for amplification. This ability to exploit chance by controlled selection will be brought out more fully in Chapters 7 and 8.

But let me say again that nothing in this chapter need be taken as denying the immense heuristic value of the "faith in the existence of causes". The breaking of the nail and the sliding of the rock certainly have their causes; and this is not in the least inconsistent with the triggering of the event being due to the amplification of chance occurrences. 'Chance' often shows itself much more clearly in regard to precisely *when* an event will occur than in regard to *whether* it will occur. For although the rock was always liable to slide, the exact instant at which this takes place is unpredictable since it may well depend on random micro-events in a contact area of atomic dimensions.

* For more detailed discussion of such phenomena see Temperley (1956) and Kelly (1973).

** It may be noted that the regressive form of Zeno's Dichotomy paradox is here avoided through the existence of 'atomic' events.

Distinction needs to be drawn therefore between the useful (if not particularly scientific) notion of causation, on the one hand, and the unwarranted belief in ontological determinism, on the other. The experience of the present century has gone to show that science no longer has the need for its 'laws' to be cast in the deterministic mould. Indeed that would be unduly restrictive; for whereas strict determinism offers only the probalities 0 and 1, statistical theory allows the use of probabilities over the whole range.

Let me end this chapter with how Popper (1974) expresses the matter: "Though we must, I think, be metaphysical *indeterminists*, methodologically we should still search for deterministic or causal laws – except where the problems to be solved are themselves of a probabilistic character."

Chapter 6

Thermodynamics and The Temporal Asymmetries

§ 1. Introduction. The possibility of establishing an objective$_2$ criterion of 'later than' was discussed in a preliminary fashion in § 3.4 and 3.5. It will be recalled that a standard irreversible process was invoked, and that fluctuation phenomena were disregarded for reason of simplicity. The status of 'time's direction' must now be considered much more carefully. Two of the key questions are whether the supposed anisotropy of the temporal order can be based on physical science, and, if so, whether the anisotropy is cosmically pervasive and applies to *all* epochs.

Let me remind the reader that the question whether or not the temporal order is anisotropic is the question whether the temporal order is a serial order like that of the real numbers, or whether it is a serial order like that of the points of a straight line. A line can be given a serial order by *extrinsic* application of the asymmetric relation 'to the right of'. As was said in § 2.4, a 1*D* being who lived in the line presumably could not tell the one direction from the other; the line's seriality is *not intrinsic*, as it is in the series of real numbers (where the asymmetry is provided by the internal characteristic* of 'being greater than'), but depends entirely on the external viewpoint. On the other hand, we who live in a 1*D* temporal order find that our consciousness does distinguish between the one direction and the other. Therefore it is quite irrelevant that we are not able to get outside of time and look at it from a different external viewpoint, so to speak. As in the case of the real numbers, the two directions of the temporal order is anisotropic is the question whether the temporal order is a now have to ask is whether this anisotropy depends solely on consciousness, or whether it is a feature of the external world.

Let's say at the outset that there is little likelihood that the anisotropy can be given a foundation at the atomic level. A vast literature developed round the various forms of the *H*-theorem, but none of it supported the early hopes that the theorem would display an intrinsic distinction between 'time forward' and 'time backward'. Furthermore the experimental studies of Brownian motion, and of other fluctuation phenomena, have shown that

---

* But see a footnote to § 2.4 (p. 23).

there are indeed events at a sufficiently microscopic level which do not provide a self-consistent ordering in regard to 'earlier than' and 'later than'. Of course the phenomenon of neutral *K* meson decay stands out as something peculiar. But apart from that all the evidence seems to support Reichenbach's view (1956) that temporal directionality cannot be based on elementary phenomena.

Nevertheless it remains true that science in its ordinary day-to-day affairs – which is to say when it is not being philosophical – tacitly accepts that there are several temporal asymmetries and furthermore that they are consistent with each other. One of them, as has been said, is provided by consciousness and this is objective$_1$ since ordering by 'later than' can be publicly agreed. Newton, we are told, saw the apple falling to the ground, not rising; and no one seems to dispute that apples do in fact fall. The practical business of science could hardly be carried on without such judgments.

Other familiar temporal asymmetries are those which relate to cause and effect, and to the existence of records of the past but not of the future. Since these several criteria give rise to a consistent ordering of events (or at least along the same world-line), this results in the strong conviction that time is indeed distinct from the space coordinates in having a 'direction'. Eddington, as is well-known, coined the term 'time's arrow' and linked it with entropy increase as a more basic criterion. In a very interesting passage (1935;86) he remarked that if entropy does not reveal an arrow no other statistical criterion can do so; for if there is thermodynamic equilibrium there is also, he said, detailed balancing with the result that direct and reverse processes of any type occur with equal frequency.

However since Eddington's time the issue concerning the arrow has been subjected to very searching philosophical and scientific scrutiny. The more carefully its has been examined the more complicated the issue has seemed to become. This has led Earman (1974) and Sklar to express doubts about the very usefulness of a search for a physical theory of time's direction. The epistemic motivation behind the search, writes Sklar (1974;355) is the assumption "that the ordering of the events in time is not something that can be known immediately or directly. Instead, there must be another relational feature holding between the events that we 'observe' and which is the source of our attribution of temporal priority". This he rejects as an adequate motivation since he regards the whole purpose of the search as unnecessary: "we know", he says (1974;402), "independently of our knowledge of the lawlike behaviour of physical processes in time, what the actual time order of events really is."

Of course I agree that we can know the order of events immediately and directly. Indeed in § 3.5 I went further and contended that our direct awareness of 'later than' has a certain logical priority over any order which might be provided by science. Yet Earman and Sklar seem to overlook that the motivation behind the search is the desire for a criterion (if it exists) which would be applicable even if man, with his conscious awareness, were not present in the world.

Before coming to the substance of the matter it will be useful to clear the decks by referring, if only briefly, to certain side issues.

Suppose it had been successfully demonstrated in Chapter 4 that 'the present' is objective$_2$, and further that 'past' and 'future' are objectively distinct. This would have been one way of establishing temporal anisotropy. However, that was *not* demonstrated in Chapter 4 and indeed there seemed to be no conclusive arguments, one way or the other. Evidently we must look elsewhere for a fully objective anisotropy.

Even so it is important to notice the extent to which, even within science and with reference to periods remote from our own lives, we use 'the present' as a reference moment for the purpose of *deducing* a relationship of 'later than'. Think for instance of two remotely past events, A and B, such as the formation of the Devonian rocks and the first appearance of hominids respectively. We don't establish *directly* that B occurred later than A. Rather do we use inferential methods to show that B occurred m million years ago, and that A occurred $m+n$ million years ago, where m and n are positive numbers. These are durations *relative to the present,* and it is then a deduction that B occurred later than A. Similar considerations apply to the predictions of future events; say to the next eclipse of the sun and to the next appearance of Halley's comet. Here too the one event would be established as being later than the other in terms of numbers of years relative to the present.

The usefulness of these remarks is that they serve to underline the extent to which objective$_1$ time continues to be the dominant time concept even within science. Our minds do not work in a reversible way (more on this in Chapter 8) and only one of the two directions along the temporal order appears to be accessible to us. As a consequence our language is deeply impregnated with words and tenses which already take the anisotropy for granted; for instance when we say "the door is opening", or when we regard ourselves as 'witnessing' or 'observing' certain events or processes. In all such usages the temporal inverses of acts of perception or cognition are tacitly excluded. The notion of an asymmetric temporal order is thus embedded in the very words through which we claim to have knowledge.

Another side-issue, already referred to at the end of § 3.4, is the question whether one should speak about a direction *of* time, or about a direction (or directions) *in* time.

It may be granted that if processes such as the spontaneous 'mixing'* of gases, the 'equalization' of temperatures, etc., were observed to occur, relative to our own sense of later than, on one occasion in one direction and on another occasion in the reverse direction, we would not regard these processes as having much bearing on 'time itself'. But of course this is not the case for they are always observed – at least in samples of material of macroscopic size – as occurring in the same direction. Furthermore the

* As in Ch. 2, words in which a temporal direction is already implicit will often be drawn attention to with single inverted commas, although these will also be used for certain other purposes.

directionality of all of these processes can be brought together under the common umbrella of entropy increase. This consideration suggests – but no more than suggests as we shall see later – that entropy increase is an aspect of the universe's history. If so there would seem little point, as was said in § 3.4, in not identifying this history with the direction *of* time.

On the other hand there are a number of authors, seized of relativity and of the *t*-invariance of the basic theories, who regard space-time as a sort of arena within which events occur. They prefer to speak of the direction of events or processes *in* time. They accept that there is temporal orientability, based on the conceptual transport of time-like vectors within a space-time manifold, but nevertheless they regard physical processes as saying very little about 'time's direction'. Earman for instance (1974;22) maintains that, if space-time is assumed to be temporally orientable, continuous time-like transport takes precedence over any method based on entropy or the like.

In view of this conflict of opinion it seems best to leave 'time' out of it (it's an ill-defined concept anyway!) and to speak instead of the anisotropy, or of the asymmetry, of *the temporal order*. * If in what follows I shall sometimes refer, for purposes of brevity, to 'time's direction' it is to be understood in this sense.

The foregoing leads on to a more substantial point. Suppose there were a kind of universe consisting of nothing but a clock. There would be no events other than the coincidences appearing on the clock face. Thus no independent criterion would be available by which it could be said that the clock had, or had not, reversed itself, ** or whether its hands were moving clockwise or anticlockwise. For there is nothing by which it could be decided what is to be counted as 'the next' number shown by the clock, rather than 'the preceding' one. Considerations of this sort show very clearly that it is meaningless to speak of a temporal direction except when there are *at least two* distinguishable processes, one of which is deemed *not* to undergo any reversals and is then used as a reference process for all the others. (Of course a criterion of simultaneity of *states* within the various processes is also required. No doubt this could be provided by the ciné apparatus of § 3.4 in most laboratory situations, but otherwise relativity theory would need to be applied.)

§ 2. T-Invariant Theories. In the literature much emphasis has been given to the point that physical theories indicate no temporal anisotropy since the most fundamental theories are all *t*-invariant. This characteristic was, of course, already present in Newtonian mechanics; *dt* occurred in the

---

* For further comments see Grünbaum (1973;788 ff). It may be added that Bunge (1968) uses the terms 'anisotropy' and 'asymmetry' in a much broader sense as meaning nothing more than that temporal duration is an oriented interval. The effect of this is unfortunate, since it loses the present meaning.

** Perhaps it might be claimed that a putative reversal requires an external cause and this cannot occur in such a universe. Yet this argument begs the issue since to use causal talk is already to assume a temporal direction.

equations as a square and therefore it made no difference whether t was taken as positive or negative. T-invariance was continued into the development of electro-magnetism, relativity and quantum theory.** It seemed justified on the ground that processes such as collisions, quantum transitions etc., as displayed by one or a few particles, appear always to have precisely the same probability of occurrence whether they occur from a state S_1 to a state S_2, in a temporal interval Δt, or from the 'time inverted' state of S_2 to the time inverted state of S_1 in the same temporal interval.

Whether or not quantum mechanics is fully t-invariant has been called into question by some authors*, particularly in relation to the measurement problem. Also there have been indications from experiment that not all elementary phenomena are t-invariant. Even so there remains a strong tradition to the effect that theories, in some sense, *ought to* display the t-coordinate as fully symmetrical and should therefore be constructed as t-invariant.

It is a good question whether this would have been the case but for the pattern which had been established by the immense success of celestial mechanics. For in fact the vast majority of phenomena we meet in everyday life are not reversible! If Newton had taken the study of irreversible phenomena as far as he took the study of planetary orbits, the subsequent development of physical theory might have taken a very different course. In the event thermodynamics arrived on the scence very late. By the time of the 19th century, when it did arrive, there was already an entrenched position in physics to the effect that reversible phenomena provide a sort of norm or conceptual ideal.

Before going further the notion of t-invariance needs to be expressed a little more carefully. As noted above it arises from the view that theory should make no distinction between 'time forward' and 'time backward', or between past and future. As such, t-invariance always requires the interchangeability of t and $-t$ in the relevant equations but it may also require an inversion of other variables as well. For instance if it is a question of the motion and collision of particles, the vectors describing the velocities of all particles must be replaced by the opposite pointing vectors. In other instances the time-inverted states of S_1 and S_2, as referred to above, may also require the reversal of a magnetic field, if it is present, or of particle spins, in order that the motions in question shall trace out the reverse paths.

Notice that the terms 't-invariance' and 't-noninvariance' refer to a characteristic, not of processes in nature, but of *theories.* On the other hand the processes themselves are said to be either reversible or irreversible, where reversibility means (as in thermodynamics) the possibility of completely restoring the original state of system *and* environment. Swin-

---

* See for example Watanabe (1966;557), the articles by Frisch, Bastin and von Weizsäcker in Bastin (1971), and Belinfante (1975).

** For a more adequate account of the t-invariance of these theories see Mehlberg (1980).

burne (1977) remarks that we have no ground for saying that theories are t-invariant or t-noninvariant until we know the difference between past and future. This is perhaps to go a little too far. But it follows from § 1 that any actual process P cannot be said to be in accordance with either t-invariant or t-noninvariant theory until one adopts a reference process R, deemed not to undergo reversals, against which P may be compared. This condition usually goes unnoticed since for R we rely so implicitly on our own sense of 'time's direction'.

Let me add that if a theory were to be t-noninvariant it would describe irreversible processes, and equivalently reversible processes require a t-invariant theory for their description. On the other hand if a theory is t-invariant the processes it describes *may or may not* be capable of occurring in either direction; in other words the t-invariance of a theory is a necessary but not a sufficient condition for the reversibility of a process which it purports to describe. There are other necessary conditions; in particular (a) the absence of disturbances due to the environment and not allowed for by the theory, and (b) the occurrence of appropriate 'initial' or boundary conditions. Let us take these in turn.

(a) *The environment.* The effect of external influences is well illustrated by one of the arguments which has been put forward against Loschmidt's objection to the H-theorem. Suppose a gas, in 'initial' state S_1, proceeds from that state to 'expand' into a vacuum and thus reaches a 'final' state S_2. At a particular instant t_2', referring to this final macroscopic state, the molecules have velocities v_1, v_2, ... v_n. According to Loschmidt's objection the inverted state $\bar{S}_2$, having the inverse vectors $-v_1$, $-v_2$, ... $-v_n$, should have exactly the same probability of occurrence as S_2 and therefore, if the H-theorem were not fallacious, we could observe a gas '*withdrawing*' to the state $\bar{S}_1$, which is the velocity inverse of S_1, just as often as we observe a gas 'expanding' into a vacuum.

This is a persuasive argument but it admits of a ready counter-objection based on the fact that the inverse velocities $-v_1$, $-v_2$, etc. do not allow of anything more than a merely momentary tendency towards withdrawal. For the process to continue towards a complete withdrawal the molecules of the containing vessel would also need to have exactly inverted motions, as otherwise the collisions of the gas molecules with the walls would not be the temporal mirror images of the original ones. However exceedingly minute external influences, acting on the walls as well as on the gas itself, would be sufficient to destroy the required correlation of the innumerable inverted velocities lasting long enough. This is nicely illustrated by Borel's example, as quoted in § 5.5, of the effect on molecular velocities here on Earth of a small shift of material on Sirius. Thus, even though $\bar{S}_2$ is as probable as S_2 itself, it is exceedingly unlikely that $\bar{S}_2$ will be followed, after the first few molecular collisions, by *further successive sets* of exactly inverted velocities sufficient to enable the gas to withdraw to an observable degree. In short, due to random external influences, the velocities won't stay inverted in the right way for a long enough duration. Although a vast variety of successive sets of velocities allow a gas to expand into a vacuum, only a very special 'choice' of successive sets of inverted velocities will result in a gas retreating to its original volume.

(b) *Initial conditions.* It is no less true for being a truism that a theory is quite insufficient to foretell what will actually happen. Over and above the theoretical equations which express the tendencies towards change within a system, one also has to use a set of initial or boundary conditions which define some 'given' state of the system. These initial or boundary conditions are no doubt adventitious and contingent whereas the theoretical equations enjoy the status of being nomological. Nevertheless the former are just as essential for purposes of prediction.

This has an important bearing on an acute difference of opinion between Mehlberg and Grünbaum concerning the anisotropy of the temporal order. Mehlberg (1961) pins his faith on the *t*-invariance of the physical theories and argues that no consideration of boundary conditions, no merely local or regional anisotropy, can have any bearing on time's pervasive isotropy as indicated by these basic theories. Therefore he believes: "... on presently available scientific evidence time should be considered as having no arrow or unique direction, and as involving no intrinsic (observer-independent) distinction between past and future. ... the only plausible way of accounting for the fact that so many well-established and comprehensive laws of nature somehow conceal time's arrow from us is simply to admit that there is nothing to conceal. Time has no arrow."

Against precisely the same background of physical theory Grünbaum (1973) argues quite differently. He asks by what right does Mehlberg assume that the time-reversible laws, as they are known to us in our limited sample of the universe, have a general cosmic relevance; our warrant, he suggests, for a cosmic extrapolation of the time-symmetric laws is certainly no greater than for a corresponding extrapolation of the factlike conditions making for observed irreversibility. "... what is decisive for the anisotropy of time is not whether the non-existence of the temporal inverses of certain processes is factlike or lawlike; instead, what is relevant for temporal anisotropy is whether the required inverses do actually ever occur or not, whatever the reason." And he goes on to argue that the non-occurrence of certain inverse processes, due to 'initial' or boundary conditions, does indeed provide a reliable indication of time's anisotropy.

My own views, as will be seen shortly, are much closer to Grünbaum's than to Mehlberg's. During the last two decades there has been a much reduced emphasis in the scientific literature on the idea of temporal symmetry, and this, if I am not mistaken, has been due to the comparative success of the big-bang model which has provided a natural 'initial' condition from which all subsequent irreversibility within the cosmos can reasonably be traced.

That initial conditions are vitally important can be seen from an example. Imagine a molecular beam in which all the particles are moving at the same speed in exactly parallel paths; imagine too that the beam is reflected by a perfectly flat surface normal to the beam's direction. Under these very special circumstances the particles may be expected to return to their starting point. However if there is the slightest deviation from parallelism collisions will occur and the beam as a whole will not return. In fact parallelism and equality of the particle's velocities cannot be ensured because of Heisenberg's Principle. Furthermore any real reflecting surface will not be flat on the molecular scale and this too will result in scattering. In general (and somewhat analogously to the previous example of a gas expanding into a vacuum) only very special, and unreal, initial conditions will result in reversible behaviour; an overwhelmingly large number of the physically possible sets of initial conditions will give rise to irreversibility. Thus, although the t-invariance of the basic theories does not in itself provide any support to the view that the temporal order is anisotropic, *neither does it exclude it.* The matter has to be examined on its merits, using empirical knowledge as well as theory, and this will be my objective in the following sections.

But first a comment on theories in relation to 'laws of nature'. Theories and laws are often conflated by philosophers and the effect is to denigrate those empirical generalisations which the practising scientists, like myself, usually call 'laws' – for example Boyle's Law, Ohm's Law and also the Second Law of Thermodynamics. It is true, of course, that theories provide a much more comprehensive understanding than do empirical generalisations; their basis in conjecture and hypothesis makes them the real pacemakers of scientific advance. But it is precisely because theories are conjectural that many of the 'laws' will continue to be regarded as highly reliable long after some particular theory has been replaced by another.

It is an aspect of the same conflation that philosophers often express a very *absolute* view of what a 'law of nature' needs to be. For instance Earman (1974;37) says of the Second Law that this "so-called law" is a false law-like statement. In a similar vein Sklar (1974;405) quotes with approval the view of theorists who "all agree that there is in fact no such true lawlike assertion and that the thermodynamic concept of entropy is vitiated in its use for describing the world by using a concept definable only in a theory known to be incorrect."* Of course I agree that as soon as the Second Law is 'explained', by use of a more comprehensive theory, it is found to be probabilistic (as are many other 'laws'). Perhaps the Second Law is just not strong enough to bear the weight of such an imposing concept as 'The Direction of Time'. Maybe some less ambitious notion of anisotropy is needed. Yet no one, as far as I know, has ever reported in the scientific literature on having observed a macroscopic process which is contrary to the Second Law!

T-invariant theory, when considered on its own, leads to Mehlberg's view – i.e. that temporal anisotropy is nothing more than a peculiarity of certain *processes,* as observed in limited regions of the universe or for

---

* Which incorrect theory is he referring to, I wonder? Thermodynamic entropy is defined by the relation $dS = (dq/T)_{\text{rev.}}$, whereas the several quantities obtainable from statistical mechanical theory should properly be regarded as entropy *analogues* since no one of them has been shown to be strictly identical with the originally defined thermodynamic entropy.

limited durations of time. Yet this view can be stood on its head. For it can be contended equally well that t-invariance is nothing more than a peculiarity of certain *theories.* These are humanly constructed and, as such, their feature of t-invariance could well be an idealisation which overlooks some important aspects of reality. Let's turn then to a careful examination of the actual evidence.

§ 3. Does Entropy Change Demonstrate Temporal Anisotropy? On the face of it processes such as the mixing of gases, the equalization of temperatures and the combustion of fuels all seem to indicate very clearly that the one direction of the temporal order is indeed distinguishable from the reverse direction.

Furthermore, according to the analysis carried out in classical thermodynamics, all such processes are irreversible *in toto* – i.e. when all consequential changes in the environment are included. Although it is true of course that gases can be unmixed, such reversals do not take place of their own accord but only by action from outside the system in question. Some external body is thereby left in a changed state, and in particular with an increased entropy. If this second system were to be restored to its original state, there would have to be entropy increase in a third system – and so on indefinitely. Irreversibility, in the thermodynamic sense, means that the total entropy must always increase in the direction towards the future as we experience it.* As Eddington remarked, the Second Law is well summed up by the story of Humpty Dumpty.

So far so good, but as soon as one accepts the atomic theory of matter it has to be concluded that the Second Law is only a probabilistic, and not an absolute, law and that occasions *must* arise when anti-Second Law phenomena will be experienced in the direction towards the future. Think for instance of the expansion of a gas into a vacuum, as discussed already in connection with Loschmidt's paradox. According to the atomic theory, the exact and continued correlation of inverted velocities which is required for a partial, or even for a complete, withdrawal of the gas is bound to occur *sometime* – even if only after countless billions of years. This would be an anti-Second Law phenomenon since it results in a reduction of the entropy of the gas without any compensating increase elsewhere.

Considerations of this sort, as analysed in detail by the Ehrenfests (1959), lead to the conclusion that the entropic behaviour of any system which is isolated for an *immense* duration *must be temporally symmetric.* 'Downgrades' of entropy must occur, during that vast duration, just as frequently as 'upgrades'. The graph of entropy against time must be the same whether we take the direction of positive time, together with the notion of 'occurring', as being towards the future *or* the past as these are to

---

* I shall not deal with the alleged subjectivity of entropy itself, as distinct from the time direction. Although some of the statistical analogues may well contain subjective elements, this in my view does not apply to thermodynamic entropy whose change depends on fully objective variables (Denbigh, 1981).

us. It is precisely because the theories used in the derivation of, say, the H-theorem are t-invariant that the long-term behaviour has to be time-symmetric. One cannot discover irreversibility, said Max Born, where none exists.

In this situation it would surely be quite unreasonable to set about demolishing the atomic theory! Although in the previous section I made some provocative remarks about theories in relation to laws, the atomic theory is much too secure for it to be set aside. Moreover very small scale fluctuation phenomena can be observed experimentally, and it is only an extrapoloation from such observations to suppose that macroscopic fluctuations also occur, although exceedingly rarely.

Neither would it be reasonable to appeal to the 'thermodynamic limit' $N \to \infty$, $V \to \infty$, where N and V denote the number of molecules and the volume respectively of the system in question. The entire universe may have only finite N and V, and, if so, theory suggests that it undergoes regressions during periods of time which are almost inconceivably long. No doubt there is a different kind of thermodynamic limit which might be appealed to, although it is speculative. Statistical mechanics must perforce reckon the number, Ω, of complexions of a system relative to the atomic and molecular quantum states; but if more were known about the nucleus it would be possible instead to reckon Ω relative to nuclear states – and of course it would then be a vastly larger number. Suppose that below the quarks there are further levels in *infinite* number; a possibility which Bohm (1957) suggested would be entirely consistent with the history of recent experimental research. If so, the enumeration of the complexions or microstates of a system would never come to an end and the time required for the exact recurrence of any one of them would presumably become infinite.

Yet this is not a profitable line of thought in regard to the problem of temporal anisotropy. For even if it is true, on Bohm's picture, that the recurrence time becomes infinite, the system in question will still display local fluctuation phenomena and the fluctuations will be temporally symmetric in regard to macroscopic quantities such as entropy. That is to say if we consider any permanently isolated system at a time t_0, and if we make *the same* statistical assumptions about times earlier than t_0 as we make about times later, the entropy curve of the system will be symmetrical about t_0.

How then can we use entropy, if at all, in connection with time's anisotropy? The method adopted by Reichenbach (1956) and Grünbaum (1973) was *not* to make the same assumptions about the past of a system as are made about its future. This is the theory of *branch systems* – systems which are not permanently isolated, due to 'branching off' from their environment. For these an *a*symmetric probability judgment can be justified because the 'past' of such systems is different from their 'future'; they were not isolated *before* a moment t_0 but are made to be isolated for times *after* t_0.

Perhaps this use of the notions of past and future, of before and after, may seem to beg the question, but I will come to that shortly. My immediate aim is to express the essential content of Reichenbach's theory, although in a rather different manner.

Suppose Jones enters a laboratory and finds there, at a time t, a system consisting of a thermally insulated box containing two blocks of metal connected by a metal rod, together with some means of measuring the temperatures of the blocks which does not significantly affect the thermal insulation. If Jones finds at the time t that the blocks differ in temperature by ΔT, what conclusion might he draw about the past state of the system?

Let's suppose he uses a *symmetric* probability assessment. Since his probability judgment towards the future is that the temperature difference will become less than ΔT, he assumes that, prior to his entering the laboratory, the temperature difference was *also* less than ΔT – and that the system had a correspondingly *greater* entropy than it has at the time t. In other words Jones assumes that the blocks had reached the temperature difference ΔT as the result of a spontaneous fluctuation from a higher entropy state such as theory tells us can indeed occur in an isolated system. Perhaps Jones had been unduly influenced by the rigour of the Ehrenfest argument about the complete symmetry of the entropy graph! Nevertheless we would surely regard him as not having made a very sensible judgment. For we, who are sensible, all know that fluctuations of that magnitude are so improbable that it is highly unlikely that one will be observed in billions of years. The conclusion we think he should draw is a very different one – namely that someone had 'prepared' * the system in such a way that the blocks of metal had an even greater temperature difference, and a lower entropy, at times earlier than t. Smaller temperature differences and higher entropies will occur later, we think, but could not have occurred earlier. (See figure overpage.)

Let's leave Jones and his incredible assumption about the past. Those of us who normally use an *asymmetric* probability judgment in this kind of situation do so as the result of experience of the-world-as-it-is. We learn to form these judgments, not because of some intrinsic asymmetry within probability theory itself, but rather because we live in a world which has every appearance of being temporally asymmetric.

This leads on to another important point which arises from our supposition that someone had indeed *prepared* the blocks of metal. We ask: How is it ever possible to prepare a system in a non-equilibrium state? And further: What makes it possible that we can witness massively irreversible phenomena outside the laboratory, phenomena such as icebergs 'melting', forrests 'catching fire' and so on? In answer to these questions it has to be accepted, I think, (*pace* Landau and Lifschitz (1969) who take a rather different view) that the world we live in is not in an equilibrium state and that it contains within itself vast stores of potentiality for

---

* Notice that the word 'prepared', like many of the other words I am using, already implies a temporal direction. For the moment this must be taken as being based on consciousness. Notice too that whenever we prepare a system, our interest is in events or processes taking place *after* the act of preparation, and not before it.

irreversible change. The possibility of even the simplest preparatory actions can be traced back, through stages, to this pervasive non-equilibrium state on the cosmic scale. The human body needs food and hence photosynthesis and hence also the radiation from the sun. And apart from what is needed for our own activity, if we are to create temperature differences (say in blocks of metal) we need sources of available energy such as coal and oil and uranium which again depend on the existence of non-equilibrium states from which free energy can be withdrawn. The presence of ice-bergs melting in warmer seas, the existence of forests, etc.,

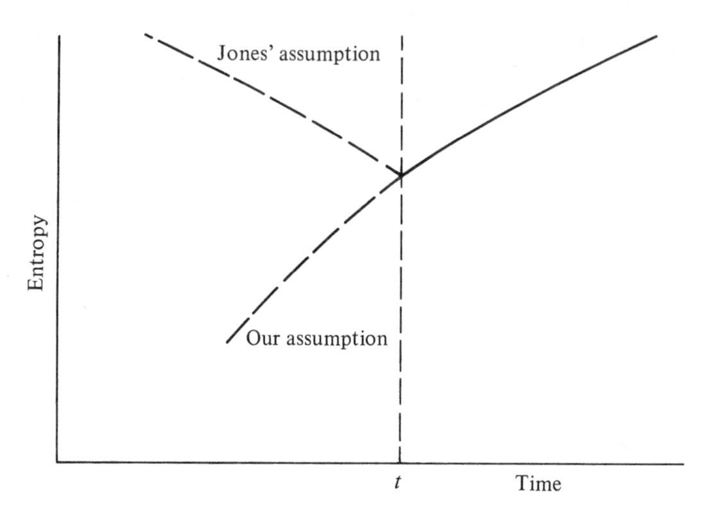

can also be traced back to the absence of thermal equilibrium. And of course astronomy reveals that there are innumerable suns in innumerable galaxies, all pouring out radiant energy arising from the irreversible conversion of hydrogen into helium.

In short the asymmetry of our probability judgments towards past and future is entirely justified by the fact we live within an evolving universe. Therefore it is highly unreal to confine the discussion of 'time's direction', as has been done by many authors, to systems which supposedly have been isolated for immense durations. Rather than seeking to devise new H-theorems which might magically display asymmetric entropy increase in isolated systems, it is far more profitable to talk about systems which are separated off from their environment at some initial instant t_0 and then, but only then, become isolated. The fact that they are separated off from a *non-equilibrium* environment ensures that the vast majority of them will also be in non-equilibrium states – and states moreover which may be expected to be random in regard to the distribution of the molecular velocities. (Or putting it more generally, the microstates of the systems may be expected to be random samples from within the set of all possible microstates)

Boltzmann had already perceived the significance of these considerations at the close of the last century. The one-sidedness of the change of H, he says, (1964;442), "lies uniquely and solely in the initial conditions". This point of view has now been widely accepted; the anisotropy of the temporal order, it is widely agreed, is to be understood as being *de facto* rather than nomological.

Reichenbach's contribution to the furthering of this understanding consisted in his study of the statistics of an ensemble of branch systems. No single one of them is really sufficient for giving expression to temporal anisotropy since there is always the remote possibility that it may undergo an anti-Second Law fluctuation. Instead, he contended, one should consider a space-average over an ensemble, thereby reducing to insignificance the effect of any one of them which might be momentarily anomalous. Indeed Reichenbach goes so far as to claim that the entropic behaviour of an ensemble does not merely correlate with the humanly known direction of the temporal order but can be used *to define* this direction. Thus from page 127 of his book:

"**Definition.** The direction in which most thermodynamical processes in isolated systems occur is the direction of positive time."

In my view however it is a mistake to use the notion of definition in this context. As was said in § 3.5, the direction of positive time *is* what it is to consciousness since we regard the consciously experienced order of events as being incorrigible by any physical criterion. Far better, I think, to regard the entropic behaviour as being *highly correlated* with the human sense of time, and thereby as being *legitimized* as a good candidate for providing a conceptual ordering of the sequence of events if man were not present in the world. As Sklar (1974;405) remarks, it would be folly "to promote a merely *de facto* regularity *discovered* to hold in the world into an irrefutable analytic proposition."

Let me turn now to a question which was raised earlier – whether the use of terms such as 'before' and 'after', 'initial state' and 'branching off' is to beg the issue. In fact these terms were used only for cenvenience. The foregoing could have been expressed equally well in temporally reversed language (where one would speak, for instance, of *final* states and of systems *merging with* their environment) or indeed in *direction-neutral* language (Reichenbach, 1956;135). Whichever language one adopts the significant point is this: in the vast majority of systems within the ensemble, the entropy changes *are in parallel with each other.* This is Schrödinger's point as used earlier in § 3.5. Any pair of members of the ensemble may be expected with high probability (so long as the systems are not too small) to obey the equation

$$(S_{Ai}-S_{Ak})\,(S_{Bi}-S_{Bk}) \geqq 0.$$

This parallelism * applies equally well whether the systems are regarded, in the one direction of time, as 'branching off' from a non-

---

* The parallelism is here considered, as in Reichenbach, as an empirical fact. A possible theoretical basis is discussed by Krips (1971).

equilibrium environment, or, in the reverse direction, as 'rejoining' it. Because of this, Reichenbach would say, the space ensemble allows of *either* of the two directions being chosen by convention as future directed. And naturally the convention he adopts is that which agrees with the human sense.

A more substantial possible criticism of the ensemble theory lies in the question whether it does indeed offer any substantial advance on the simpler treatment I gave in § 3.5 where fluctuation phenomena were excluded from the discussion. Can it really be said that the consideration of a space ensemble, as initiated by Reichenbach and modified by Grünbaum (1973), has eliminated fluctuation effects successfully? No doubt the consideration of the average behaviour of a large number of branch systems greatly minimises the significance of any one of them which may display anti-Second Law behaviour. *Even so we are still dealing with probabilities rather than with certainties.* Indeed if one supposes the branch systems to be held isolated for immensely long periods, of the order of their Poincaré recurrence times, a great many of them may be expected to have undergone large entropy fluctuations during that period. However many members there are in the ensemble, and however large each one of them may be, it cannot be excluded that the majority of them may behave 'unreliably'.

In my view nothing which offers *certainty* in regard to temporal anisotropy can be obtained out of thermodynamics or statistical mechanics. It is far better, I think, to regard the entropic behaviour of systems as having a sort of *emergent* quality – i.e. in the sense of becoming more and more reliable the larger is the system in question (or the more numerous are the systems within an ensemble), but never providing the complete certainty which would allow one to say: "Because of this, Time has a Direction." And neither can one say: "Time is one-way only", as indeed it appears to be to consciousness.

This point about emergence can be illustrated with some calculations based on Einstein's fluctuation theory. The probability P of trapping a system, by dividing it with a shutter, in a state having a Boltzmann entropy less than the maximum by an amount greater than or equal to ΔS, is given by the equation

$$P = e^{-\Delta S/k} / (\pi \Delta S/k)^{1/2}$$

as quoted by Mayer and Mayer (1940). Take the case of helium at 273K and 1 atm. for which the translational (Sackur-Tetrode) entropy works out at $S/k = 9.0 \times 10^{24}$ per mole. The second column in the table gives values of

N	P	Years
10^6	10^{-7}	19
10^{10}	10^{-10^5}	10^{10^5}
10^{20}	$10^{-10^{15}}$	$10^{10^{15}}$
10^{24}	$10^{-10^{19}}$	$10^{10^{19}}$

the probability of trapping a system consisting of *N atoms* of helium in a state which has an entropy only *one millionth* less than the Sackur-Tetrode equilibrium figure. The third column then shows the duration one would expect to have to wait to observe that fluctuation if the shutter were momentarily closed once a minute.

The figures show very clearly that the probability of a fluctuation even as small as a millionth diminishes extremely rapidly with increasing size of the system. The 'reliability' of the Second Law may correspondingly be said to increase in proportion to the time one would have to wait to observe that one millionth deviation – some $10^{10^{19}}$ years in the case of not more than 6.6 grams of helium! * It's not surprising that astrophysicists, who normally deal with vastly larger quantities of matter, rely in their work on the Second Law with a sense of complete security.

The upshot of these remarks is that 'time's direction' as given by entropy change, as Landsberg (1972) has emphasized, is a macroscopic and probabilistic concept. There is no such 'direction' in systems consisting of only a small number of molecules. The entropic anisotropy emerges, so to say, in sufficiently large systems, and with even greater reliability in ensembles. That today I shall observe a speck of ice 'forming' in a glass of warm water is exceedingly improbable, and that I shall observe the whole glassful 'freezing' is almost inconceivably improbable. Yet, even if the unthinkable were to happen, it need not shake my confidence in the usefulness of the Second Law, or in my own awareness of the normal sequence of events. For I would still be surrounded by a vast number of other events and processes taking their accustomed course: the ink 'flowing' from my pen, the sun 'shining' through the window; and so on and so on.

Even so, entropy change cannot be regarded as an indicator of 'time's direction' throughout an infinity of time. Indeed during an infinite period fluctuations of any magnitude may be expected to occur an infinite number of times! And of course this is a further reason why it is inappropriate to use the entropic criterion as a *definition* of a direction; for if it were so used one would be driven to the conclusion, as will be seen in § 5 in connection with one of Reichenbach's conceptions, that 'time' has opposing threads.

Also it needs to be kept constantly in mind that verbs denoting things 'happening', 'forming', 'freezing', etc., (as brought out by the use of inverted commas), describe the direction of processes as this direction is to human consciousness. As was seen in Chapter 4, no strong evidence is available for choosing between the *A*- and *B*-theories. If one is nevertheless inclined to adopt the *B*-theory, as being more congenial to physics, one has to

* The time for *any one* of the Ω microstates of the gas to be repeated may be expected to be even greater. Of course this recurrence time bears no immediate relationship to the times shown in the table: (a) because the latter refer to the trapping of the system in an improbable *macro*-state, and (b) because the intervals between the momentary closings of the shutter have been arbitrarily chosen as one minute, a period appropriate in relation to the rate of relaxation of the gas. (See Huang, 1963;89.)

accept that there is no 'coming into being' of things or events; therefore, *apart from human consciousness,* there is no reason for choosing the one direction of time as being any more real than the reverse direction. Indeed, from the viewpoint of the *B*-theory the universe does not 'evolve'; it merely has different entropy values, or different degrees of expansion, at different clock times. In short, we can only speak of 'evolving', 'forming', 'freezing', etc. when the matter is looked at from the human point of view.

To be sure, when we think of the 'shuffling' of a pack of cards we tend to regard it as natural that the shuffled state of the cards is later than the sorted state, that the system 'becomes' more probable. But that is only because we inject our personal consciousness of what is 'later' and our sense of 'becoming'. We see the process as a shuffling, and not as the time inverse which is a sorting.

Thus if we wish to adopt the *B*-theory, and to use it consistently, we have to abandon all idea that there is a basic 'ongoing' in the world; all parts of time are 'equally real' and the only thing that can be said, on the basis of the Second Law, is that there is a steady *gradient* of entropy values along a world-line. As far as physical science is concerned, if it adopts the *B*-theory, the gradient can equally well be regarded as 'up' or 'down'. The notions of 'later than' and of 'becoming' are entirely mentalistic, although we can, *by convention,* denote the entropy curve as 'rising' in order to obtain correspondence with the human sense of time's ongoing.

Such a view, if it is adopted, does not discredit the foregoing account of the Second Law as arising from the vast potentiality for change which existed in the initial state of the universe. It is rather to say that it is we who regard that state as being 'initial', and that, given our sense of an ongoing, experience leads us to adopt a certain probability judgment – one which regards it as natural that the 'more probable' states of isolated systems should be later states rather than earlier ones. No doubt we would reverse that judgment if we habitually experienced pairs of bodies becoming more unequal in temperature, and similar anti-Second Law phenomena. But (as I shall argue in § 6) it would not be possible for creatures such as we are to survive under such conditions!

Finally a few words about the formulation of the Second Law. It has been mentioned already that Earman and Sklar regard this "so-called law" as a false law-like statement and Popper (1957–8) has rightly pointed out that the law is strictly falsified whenever a heavy colloidal particle, suspended in water, is momentarily lifted up due to the irregular molecular impacts. In such a process, as he says, there occurs an isothermal conversion of heat into an equivalent amount of work, contrary to the Second Law as it is normally expressed.

Of course the traditional formulations, as adopted by Clausius and Kelvin and Planck, were never intended to allow for fluctuations. Let's ask whether the law can be reformulated in a manner which will allow for them and yet retains the law's status as an 'impossibility theorem'. One such variant reads as follows: "It is impossible to predict the occurrence of a process, within any chosen interval of time, in which a withdrawal of heat from a heat bath, at a uniform temperature, results in the obtaining of work without leaving a change in the thermodynamic state of some other body." Or more simply (Jaynes, 1963): "Spontaneous decreases in the entropy, although not absolutely prohibited, cannot occur in an experimentally reproducible process."

Yet another version is due to Penrose (1970;222). He draws attention to the fact that, because fluctuations are random, they cannot be utilized by an engine operating in a regularly periodic manner; hence "There can be no cyclically deterministic perpetual motion machine of the second kind." By 'cyclically deterministic' he means that the observational state of the total system at the end of each cycle of operations is uniquely determined by its state at the beginning.

§ 4. Outwards Flow and Causality. Thermodynamic processes are of course by no means the only candidates as indicators of temporal anisotropy. A rather different kind of process is the outwards flow from a centre, accompanied by dispersion in space.

One of the simplest instances is described by Whitrow (1980;10) as being due to E. A. Milne: "He noted that any swarm of non-colliding particles moving uniformly in straight lines, if contained in a finite volume at some particular initial instant, will eventually, that is at some finite later time, be an expanding system, even if it were originally a contracting one. On the other hand, an expanding system of uniformly moving particles will never of its own accord become a contracting one. Although Milne considered a swarm of particles (with a cosmological analogy in mind), it is sufficient for our purposes to consider only two particles. If initially they are approaching each other, then eventually they will be found to be moving apart. But if initially they are moving apart, they will continue to move apart and will never approach each other."

Milne's example is clearly similar to that of a gas released into empty space, as discussed by Popper (1956b) and Grünbaum. Although the actual experiment has perhaps never been done, one feels quite confident that if a phial of gas were to be broken in empty space the gas molecules would stream away in all directions to the ends of the universe; one would certainly not expect the reverse phenomenon to occur, one in which the molecules spontaneously arrive and reassemble themselves in the broken phial. Just why we feel this confidence is due, I think, to intuitive probability considerations fortified by the knowledge that a gas does indeed expand into the whole space of a *closed* vessel. But whereas the latter process is known to be entropy creating, there is a certain technical difficulty in the way of calculating what may be the entropy change of a gas expanding into the whole universe. Nevertheless it can surely be said on intuitive grounds that the probability of each one of a number of particles being *anywhere* in space is vastly greater, other things being equal, than the probability of their being close together.

Another familiar example concerns radiation. If one sets up a glowing filament or an oscillating dipole, its radiation *goes outwards* and continues to do so until it is absorbed. The time-inverted process is not observed; relative to *our* time direction a cold filament does not become glowing, nor is a dipole set oscillating, by the action of inwards converging radiation. Of course if radiation is regarded as consisting of photons this example is much the same as that of the outwards flow of gas into space; on the other hand if radiation is regarded as a wave process the analogy is weaker and we seem to be dealing with something rather different. As is well known Maxwell's theory of electromagnetism is t-invariant and thus, within the scope of the theory, outwards and inwards moving waves are both equally permissible. In other words the Maxwellian equations allow of *two* solutions; one of them is referred to as the retarded field solution and the other as the advanced field solution. In the application of the theory to 'real life' situations the advanced field is usually disregarded as being 'non-physical', and only the retarded solution, which corresponds to outgoing waves, is taken as physically real. *

Yet another example of the apparent irreversibility of outwards flow from a centre was pointed out by Popper in a series of short papers (1956–67). This is the simple example of circular waves moving outwards to the periphery of a pond as the result of a disturbance near its centre. A film of this process run backwards would display the 'non-physical' process of waves being initiated at the pond's periphery and subsequently converging to a central point.

Popper's papers gave rise to considerable discussion on the possibility of the existence of *non-entropic* indicators of temporal anisotropy. This indeed was Popper's objective; for although the water waves are damped through viscous action (an entropy creating process), he rightly pointed out that this is an adventitious factor since one can readily conceive wave motion on an idealised non-viscous medium. His purpose, he said, was to confute the widespread belief that the 'arrow of time' is solely dependent on the Second Law. Hence the significance of his pond example, and it will be noted that it differs from the particle examples, as previously quoted, in two further respects: (a) the water molecules do not move outwards, and instead there is a radial flow of energy; (b) *coherent* waves would have to be produced at the pond's periphery, for the reverse process to occur, and this may seem to render this reverse process as doubly improbable. The water waves, as Popper remarked, could only be seen as moving inwards to the centre if the entire periphery of the pond contained a set of *coherent* wave generators acting as an "inexplicable conspiracy of causally unrelated conspirators."

Now the factual irreversibility of the foregoing group of processes can hardly be gainsaid, but the question I am turning to is how this irreversibility is to be understood. For it is only through understanding its

* Inwards moving radiation can of course be obtained by use of a spherical mirror. However, this radiation would be expected to have originated, in the first place, at the centre of curvature – e.g. at a central filament – and to have been reflected at the mirror only at a later time.

origin that one can form a judgment about whether 'outwards flow from a centre' is in any way *more absolute,* as an indicator of temporal anisotropy, than is entropy increase. Let's ask then whether it is any *less* probabilistic. To be sure Popper, and some other writers on the subject, have expressed their own understanding of this kind of irreversibility in terms of causality, and I will come to that later. First let's see how far one could go on the same probabilistic lines as are used in the molecular interpretation of the Second Law.

We first recall that thermodynamic irreversibility can be made understandable * by distinguishing between macrostates and microstates (or complexions), and by noticing that certain macrostates may comprise a vastly larger number, Ω, of microstates than others. If the individual microstates are all equally probable (as is generally assumed in the case of a system at constant energy), the macrostates of high Ω become thereby immensely more probable than those of relatively low Ω. Therefore if a physico-chemical system has been prepared in a non-equilibrium state, its process of change will most likely be towards a macrostate of higher Ω. In the Boltzmann-Planck interpretation, which is sufficient for present purposes, entropy increase is taken as the increase in the value of $k \log \Omega$**, and the reverse change is not impossible but exceedingly improbable. It will be seen that causal ideas do not enter into this account at all significantly.

Consider now the various examples of 'outwards flow from a centre'. The case of radiation within a relatively cold *enclosure* is, of course, a straightforward instance of thermodynamic irreversibility. There occurs a transfer of energy from the 'hot' filament or dipole to the cold surroundings with increase of entropy.× On the other hand, as mentioned above, there are certain technical difficulties in treating the radiation of photons, or the expansion of a gas, into *unenclosed* space by the methods of statistical mechanics. These difficulties, which have been discussed by Grünbaum (1973) and Zenzen (1977), arise from the question whether or not a Boltzmann entropy can properly be attributed to particles in *open* space, especially if this is infinite in extent. For this reason there is some doubt about these processes being entropy-producing. Nevertheless their irreversibility is still entirely understandable on probability grounds, namely: (a)

---

* Wehrl (1978), Penrose (1979), and Prigogine (1980) give reviews of recent work on the rigorous solution of the problem of irreversibility within statistical mechanics.

** Of course the use of *the same* probability judgment towards the past results in the paradoxical conclusion that Ω, and hence the entropy, will also increase in 'time backward', as indicated in § 3.

× These remarks are entirely consistent with the absorber theory of Wheeler and Feynman (1945). These authors remark that the one-sidedness of radiation is not purely a matter of electrodynamics; it is rather a statistical matter connected with the radiating body being in a special initial state such that it is far more probable that it will lose energy than gain it. Similarly Davies (1975) points out that the cooperative emission of radiant energy by a large number of particles at just the right moment to bring about a coherent wave, converging to a centre, would be a fluctuation phenomenon which is exceedingly improbable. For a comprehensive review of absorber theory in relation to 'time's arrow' see Pegg (1975).

the 'special' initial state where particles or photons are confined in a small volume, and (b) the surely very high probability that the particles or photons will move further apart (so long as their interactions may be neglected) if a large or infinite amount of space becomes accessible to them.

Let's turn to the pond example. The normally assumed initial state is one in which a pebble is dropped in near the centre and gives rise to coherent waves moving outwards. Notice however that an entirely practical alternative is provided by the dropping of a large circular hoop. This would give rise to coherent waves moving inwards! But of course I agree with Popper that it is very improbable indeed that coherent waves would be similarly generated at the edges of the pond, since the edges do not provide a comparable mechanism. This leads us to ask: Is it not almost equally improbable that coherent motions would be generated spontaneously *at the centre* – i.e. in the absence of the mechanism provided by the pebble?

This point, about the improbability of the central generation of waves, as well as of their peripheral generation, has been largely overlooked in the literature. It leads me to the view that the entire class of processes considered in this section are probabilistic. *Ipso facto* they do not display absolute irreversibility; the reverse processes *could occur.* The question then arises: Are such reverse processes *any more* improbable than are the various kinds of thermodynamic process already discussed in § 3?

Let's recall that the probability of a mere one millionth regression in a gram mole of helium is as minute as $10^{-10^{19}}$. Can it be confidently asserted that the reversals of any of the processes described above have an even smaller likelihood of occurrence than the reversals of thermodynamic phenomena, when like is compared with like? I think not and wish to suggest that it has been the existence of a well-established fluctuation theory, in regard to the thermodynamic phenomena, which has given rise to a contrary opinion. Fluctuation theory has provided a reason for *doubting* the thermodynamic arrow. It would be quite wrong to suppose that the criteria of temporal anisotropy of the present section are necessarily more absolute than is the thermodynamic criterion *; it is simply that there is the lack of a corresponding theory for estimating the probability of the processes in question occurring in reverse.

Before proceeding to a discussion of causality, mention should be made of the contributions made by Hill and Grünbaum (1957) and Penrose and Percival (1962). The former remark that a weak point in Popper's presentation of the pond example is that it turns on the somewhat vague notion of 'spontaneity' – i.e. that the generation of coherent waves at the periphery is exceedingly unlikely to occur 'of its own accord'. In order to

* Perhaps it may be said that the expansion of a gas or of radiation into an *infinite* universe is indeed absolutely irreversible. But so also is entropy increase at the thermodynamic limit $N \to \infty$, $V \to \infty$, which is the comparable idealization. It may be hoped that a statistical treatment will be found which will be capable of bringing both processes within the compass of the same probability metric.

avoid making this awkward distinction between what may be described as spontaneous and what may not, Hill and Grünbaum focussed their attention on a supposedly infinite universe and they put it forward as a principle that, within such a universe, there exists "a class of allowed elementary processes the inverses of which are unacceptable on physical grounds by requiring a *deus ex machina* for their production". A somewhat similar principle was advanced by Penrose and Percival in their postulated 'law of conditional independence'. This is to the effect that causal influences coming from infinity in different directions are independent of one another. ** The authors gave several examples of the use of this principle for dealing with irreversible phenomena.

Turning then to the 'causal' treatment of this group of processes, distinction needs first to be made between the symmetric relation of causal connectibility, on the one hand, and the asymmetric relation between cause and effect, on the other. The former is scientifically very respectable; the latter, in my view, lacks any clear scientific warrant. However useful the terms 'cause' and 'effect' may be in everyday life, they are not scientific concepts. The more elementary are the events in question, the more the supposed distinction between 'cause' and 'effect' *seems to disappear!* Furthermore the usage of these terms presupposes the very issue which is our present concern – i.e. the order of temporal succession. Thus it just won't do to say that advanced fields must be rejected because they are contrary to causality!

Let's suppose, contrary to fact, that we *always* experienced waves or particles as coming inwards to a centre. I believe we could still give a rational account of the phenomena, using causal notions, and if so it follows that these notions are not useful for explaining why, in fact, we do not experience the supposed reverse movements.

My starting point is the acceptance of Popper's basic contention: "Only such conditions can be causally realized as can be organised from one centre . . . causes which are not centrally correlated are causally unrelated and can cooperate only by an accident (or by a miracle)." In the hypothetical situation where we experience the foregoing processes as occurring *inwards,* the centre could still be thought of as the origin of influence but it would be regarded as *drawing* the waves or particles inwards, rather than as *sending* them outwards. In short we would need to modify our present conceptual framework and, for these kinds of process at least, 'pushes' would be replaced by 'pulls'. The cause (i.e. the centre) would be regarded as acting backwards in time (*our* time) rather than forwards. We would then adopt the rule, for this group of processes, that effects precede their causes and there would seem to be nothing odd about this (except perhaps if other kinds of process had not also reversed themselves.) Furthermore the influence exerted by the centre would appear as affecting distant regions earlier than it affects nearer regions –

** In the case where the universe is assumed to 'begin' with a singularity, the authors remark that their posulate depends on the axiom that all spatially separated regions became uncorrelated from the beginning. See also Davies (1974;119).

where 'earlier' is again as it is to us. This is not in the least irrational – and indeed it is just what is contemplated when Maxwell's theory is used with advanced solutions.

Perhaps the reader will object however that there are other examples where 'reverse causality' or 'finalism' would certainly be irrational. He might instance the case of the recording barometer discussed by Reichenbach (1956;183). As we normally experience the action of the barograph, a change of atmospheric pressure is quickly followed by a movement of the pen and this, in its turn, is followed by the deposition of ink on the revolving drum of paper. Suppose however we always experienced this process in reverse; the ink trace would be seen as being progressively deleted and, at each instant, the pen would adjust itself to a point on the paper where the ink had already disappeared and this would be followed (in our time sense) by the atmosphere having the pressure which corresponds to the vanished pen mark. It would seem absurd (so goes the objection) to suppose that the vanished ink mark influences the pen to be at exactly the right position on the paper, and even more absurd to suppose that it also influences the pressure of the atmosphere to have a particular value. Yet these suppositions, as described by Reichenbach, miss the point for they are absurd only in relation to the rule that effects *follow* causes. As I see it, in the hypothetical reverse sequence of events we can continue to regard the atmosphere as the active agent (just as in the pond and gas examples we can continue to regard 'the centre' as the origin of influence), but we must think of the pen as being moved by a slightly *later* state of the atmosphere (as 'later' is to us), and similarly we must regard the deletion of the pen mark as being influenced by a slightly later position of the pen.

Yet there *is* a peculiar feature of this example and it is this: the *deletion* of the pen mark. The reversal of the mechanical processes of atmosphere and pen can be dealt with, as has been seen, by reversing the rule for the temporal order of cause and effect, but the deletion of the pen mark would be the reversal of a thermodynamic, and not of a merely mechanical, process. And indeed quite generally the truly paradoxical character of observing reverse processes would reside in the need to reverse our *probability* judgment relating to entropy, far more than in our need to reverse our *causality* judgment relating to non-entropic movements.

In view of the foregoing it is not surprising, as was said in § 5.2, that the 'arrow' which is presupposed in asymmetric causal statements appears always to be consistent with the 'arrow' provided by thermodynamics. In a sense they are made to be consistent through an understanding of what distinguishes 'cause' from 'effect'; the distinction lies solely in some part of the cause being always earlier than the earliest part of the effect. "The distinction between cause and effect", says Reichenbach, "is revealed to be a matter of entropy and to coincide with the distinction between past and future." (1956;155)

Notice too that, within the scope of a classical type of theory which is fully deterministic, there is no call for distinguishing between cause and effect. For if event e_1 is a necessary and sufficient condition of event e_2, then e_2 is a necessary and sufficient condition of e_1 and there is no logical asymmetry. As Wheeler and Feynman remark (1949;425), in deterministic dynamics "the distinction between cause and effect is pointless ... the stone hits the ground because it was dropped from a height; and equally well: the stone fell from a height because it was going to hit the ground." But, of course, this is not to say that we cannot decide *to label* states-of-affairs as cause and effect according to their temporal order.

So much for a brief survey of the phenomena of outwards flow from a centre. It has been shown that it is mistaken to 'explain' these phenomena in terms of causality. They can be made fully understandable on grounds of initial states and probability. As such they do not provide any absolute criterion of temporal anisotropy, except in the idealised case of outwards flow into a supposedly infinite universe.

Perhaps it should be added that the expansion of *radiation* from a centre has become widely known as indicating 'the electromagnetic arrow of time' (*vide* articles in Gold, 1967). In view of the closely similar processes displayed by the spreading of particles into empty space it would be preferable to use the more comprehensive term 'the dispersion arrow'. A further instance of dispersion, the expansion of the universe as a whole, will be the subject of the following section.

Let me end with a remark which has a bearing on the well-known EPR paradox in quantum mechanics. It will be recalled from § 3 that 'time's direction' was found to be a statistical concept; the direction 'lasts' for a shorter duration the smaller the number of particles there are in a given non-equilibrium system, and it cannot be said to exist at all at the level of single quantum events. This view is in accordance with the *t*-invariant theories which apply to these events, and it also implies the symmetrical application of advanced and retarded fields to unitary events of absorption and emission. It follows that the terms 'earlier than' and 'later than' are not intrinsic to single quantum events, and neither are the terms 'cause' and 'effect' since these are no longer in any way distinguishable. Thus it is a quite unnecessary mystification to speak of 'messages' in the EPR system as going 'backwards in time', or as involving a causal anomaly. At that level of the application of the time concept there is no backwards or forwards, and neither is there any causality in the sense of cause preceding effect.

§ 5. Cosmological Theories. A number of scientists hold the view that the most basic 'arrow', the one which determines all the others, is the expansion of the universe. Let me make a couple of comments on the language of the matter, before coming to its substance. The first is that the supposed 'expansion' is relative to, say, the standard metre or to the radius of the Earth's orbit, which are assumed to remain, in some sense, of constant size. No doubt an alternative assumption is that all terrestrial and atomic lengths are contracting relative to the universe, and there seems to be no way of establishing the one assumption in favour of the other – or, at least, without making further assumptions relating to, say, the constancy of time-keeping devices and the speed of light. My second comment is that to speak of an 'expansion' is to make tacit use of a temporal reference direction. It would be more correct to say that what is referred to is a divergence of the world lines of galaxies along the one direction of their proper times, and a convergence in the reverse direction.

A good deal of very speculative cosmology is involved in the attempt to 'reduce' the other arrows to the cosmological arrow. Although the latter may be more fundamental, it is certainly less well-founded! I mean in the sense that cosmological theories are much more uncertain, and are much more liable to frequent re-interpretation, than are the theories of thermodynamics and statistical mechanics which refer to ordinary man-sized lumps of matter. Indeed Alfvén (1976) rather unkindly described the current big-bang theory as myth-making, a modern form of the cosmology of the crystalline spheres.

To be sure the notion that the universe is expanding is not dependent on big-bang theory, but derives rather from a vast amount of experimental work on the red-shift of the galaxies. Yet even the conclusion that the red-shift implies expansion is not entirely secure, since some other explanation of the experimental results may be found. One such explanation which has been put forward is the 'tired light' hypothesis which supposes that space is populated by a type of particle, the φ particle, which scatters the light from distant galaxies and gives rise to a non-Doppler red-shift. Although this hypothesis seems no longer to be taken very seriously by cosmologists, I mention it to show that the expansion of the universe doesn't have the *factual* character with which one can assert that hot and cold bodies in contact do indeed equalise their temperatures, and that particles and radiation normally spread outwards from a centre.

In spite of these reservations – and a more important one will be put forward shortly – the theory of the expansion and the derived theory of the big-bang have undoubtedly helped to provide a clearer understanding of the other temporal asymmetries. This may be seen from the work of Gold (1958, 1967, 1974), Gal-Or (1974, 1975) and P. C. W. Davies (1974), although the latter gives a very different emphasis.

The scientific reader will be familiar with the general picture presented by the big-bang theory and in any case the relevant parts can be very briefly summarised. Early in the history of the universe matter and radiation were in thermal equilibrium with each other at an exceedingly high temperature. However the continued expansion of the universe resulted in its cooling and when the temperature fell to about 3000K electrons were able to combine with atomic nuclei. Consequently an insufficient number of free electrons were left in space to scatter the photons. This entailed that matter and radiation no longer interacted strongly; they became 'decoupled' and henceforth ceased to be in thermal equilibrium.

Matter then began to cool more rapidly than radiation. Yet at a still later stage this trend was reversed due to the onset of a new process – the aggregation of matter to form gaseous clouds, and eventually the galaxies and stars. This process remains very mysterious but it has been widely assumed that gravitational aggregation was initiated by local inhomogeneities, perhaps in regions where expansion was occurring more slowly than elsewhere. Be this as it may, the formation of the stars was attended by their rapid rise of temperature, due to the release of gravitational energy, and this led to the onset of thermonuclear reactions in the stellar interiors. Thus the present state of the universe is one which is much more dispersed, in a spatial sense, than it was at an earlier stage, and it also displays vast differences of temperature between the stars, on the one hand, and the radiation in space, on the other. The stellar interiors are at temperatures perhaps as high as 10^{10}K whilst the background radiation in space has by now fallen to as low as 3K.

Gold (1958) has imagined the hypothetical enclosure of a star within a box having non-conducting walls. When the star eventually reaches a state of thermal equilibrium with the internal surfaces of the box, all 'arrows' within the box will be lost. This consideration leads him to the conclusion that because empty space is so cold, due to the universal expansion, the expansion process is *the origin* of all local arrows of time, and in particular is the origin of the thermodynamic arrow. "Without the radiation sink that the universe provides," he writes (1974), "sub-units would have reached thermodynamic equilibrium and that means that no statistical anisotropy along the time axis would exist; . . ." This view has been developed further by Gal-Or (1975) who regards the time asymmetry related to the expansion of the universe as being the 'Master Arrow' which dominates all others.

As has been said, Davies (1974) gives a different emphasis. He points out that the thermonuclear reactions within the stellar interiors are irreversible processes whereas the cooling of radiation due to the universe's expansion is reversible, at least to a first approximation. If the present expansion phase were to be followed by a re-contraction, as it can be according to certain models, this would cause the background radiation temperature to rise but without in any way affecting the thermonuclear processes within the stars. For this reason "the expansion of the universe is irrelevant to the emission of starlight", and indeed the creation and emission of photons by the stars would continue long into a contracting phase (if it ever occurs) – and would continue, in fact, until the temperature of the background radiation had risen to a value at which it would be in thermal equilibrium with the stellar interiors. *

In Davies'view it is therefore the thermonuclear furnaces of the stars which constitute the main power houses of the universe at the present epoch and they are a source of dis-equilibrium far more significant than is

* Davies also shows that the expansion of the universe is insignificant in regard to Olber's paradox; what matters for the resolution of the paradox is not the continued process of expansion but rather the universe's age and the distance from which light could have travelled to the Earth within that time.

the continued expansion. Indeed the latter has remarkably little influence on *local* thermodynamics here on Earth. It is the Sun's energy which permits photosynthesis, and thereby the laying down of coal and oil deposits such as allow of the setting up of local branch systems which demonstrate entropy increase. And it is the same source of irreversibility which makes it possible to create oscillations in dipoles or to initiate wave motions on ponds.

It remains true of course (within the scope of the accepted theory) that, if the universe had not expanded in the first place, there would not be the state of dis-equilibrium which exists. To that extent the expansion *has been* (rather than *now is*) the dominant factor. This point can be further illustrated by showing the ambiguity which would arise if the expansion process were used for purposes of *defining* a 'direction of time'.

As mentioned already, some of the theoretical cosmological models allow of the expansion 'ending' and being 'followed', in some sense, by a contraction. Consider for instance the Friedmann model. As is well-known this is a highly simplified model in so far as it assumes that the universe is filled by a homogeneous and isotropic medium and thus does not allow for the existence of the stars. Nevertheless it is sufficiently simple to permit the equations of general relativity being put in a solvable form. With the further assumption that the pressure of the medium is negligible compared to its energy density, the following equations are obtained (Weinberg, 1972;472–481; Davies, 1974;82):

$$\varrho R^3 = \text{const.}$$

$$(dR/dt)^2 = \frac{8\pi G \varrho R^2}{3} - \mathrm{k}.$$

Here R is the scale factor for the distance between any two co-moving points (i.e. R is a measure of the expansion), G is the gravitational constant, ϱ is the density and k is the curvature index which has the values $+1$, 0 or -1. Consider the $k = +1$ case. The form of the equations shows that the Friedmann model with this value of k allows of R passing through a maximum value. It would appear therefore that the model universe 'starts to contract' at some stage as the time variable t 'continues to increase'.

But what do these words mean? What they must surely mean in operational terms is that there exists some process or clock which is assumed to 'continue' providing larger values of t quite undisturbed by the assumed reversal of the universe's expansion. And further that it is the process or clock (and not the universe) which is regarded as providing the criterion of 'later than'. The process or clock must be deemed not to undergo any reversals.

The point I am making is that it is precisely because there exist models of the universe which permit re-contraction (models it may be added which cosmologists regard as being fully meaningful) that it becomes futile to take the temporal direction of *expansion* as providing a *definition* of 'later than'. For if it were so defined, rather than by use of the supposed process or clock, the universe could never be said to contract: the verb 'to contract' implies being smaller at later times; but if 'later than' is taken *by definition* as always corresponding to 'larger than' there can never be any later contraction!

In everyday situations we depend implicitly on consciousness for providing the reference process which is deemed not to undergo reversal. Yet this is clearly unsatisfactory in regard to the supposed contraction. For suppose it is calculated to 'begin', by use of some estimated value of ϱ, in, say, 6×10^{10} A.D. This would be to think of the A.D. extrapolated forward to that year, with unity added to the date each January 1st. But, obviously enough, humans may no longer be present at that time and we need therefore to contemplate some alternative criterion of the temporal direction of the kinds already considered in § 3 and 4.

A long-lived entropy creating process such as radioactive decay is a suitable candidate since there is no reason to suppose that it would be affected in any way by the onset of contraction. But let us ask whether *gravitational collapse* might also be a suitable candidate. As is well known the collapse appears to have no stable end-point short of a singularity. A self-gravitating system, says Davies (1974;108), "therefore possesses an infinite reservoir of negative entropy. . . . The origin of *all* thermodynamic irreversibility in the real universe depends ultimately on gravitation." *

This may well be the case, but it can hardly be said that gravitational processes provide a really *usable* criterion of temporal anisotropy. The collapse, as it supposedly occurs in black hole formation, is not a directly observable phenomenon – not in the sense in which thermodynamic processes are directly and readily observable. We could not use the black hole process as a practical means of ordering the sequence of events, long before man existed, as we do in fact use the decay of radioactive elements. Of course what *can* be observed is that bodies fall towards the earth; the gravitational force, in non-relativistic terms, is one of attraction and not of repulsion. Yet in the case of inelastic bodies which strike the earth without bouncing back we are dealing once again with an entropy creating process! And elastic collisions are *t*-invariant.

The conclusion I think is this: although cosmology does indeed offer a deeper understanding of irreversibility, it does not provide an indicator of temporal anisotropy which is to be preferred to long-lived thermodynamic processes such as radioactive decay. Entropy increase is a generalization of a great many kinds of irreversible processes such as nuclear reactions, the absorption of radiation and the gravitational collision of inelastic bodies. Its advantage as an indicator of 'time's direction' is nicely illustrated by Landsberg and Park (1975) (See also Walstad, 1980). These authors use a form of the Friedmann model which is more realistic than the one quoted earlier in so far as it allows for dis-equilibrium between matter and radia-

* See also R. Penrose (1979) who suggests that the universe may develop from an 'initial' state having a very special geometry to a 'final' state whose geometry will be very chaotic. The growth of geometrical irregularities provides by far the largest contribution to a continually increasing entropy in Penrose's view.

tion. They show that entropy will continue to increase during a contracting phase, and furthermore that it will still continue to increase (so long as the peculiarities of the $R = 0$ states may be disregarded) throughout the duration of successive cycles of contraction and expansion. Landsberg and Park therefore claim with some justice that their model "is compatible with the view that an increase of entropy defines the advance of time, even in a contracting world model."

This brings me to a rather distinct but related topic – the notions of 'reverse worlds' and of 'closed time'. The idea of reverse worlds can be understood in two different senses:

(1) Apart from our own universe there 'co-exists' *in space* another universe having all of its processes counter-directed.
(2) Our own universe (or parts of it) undergoes reversals *in time.* These reversals are either partial or complete, where in the latter case it is meant that they proceed to entirely identical states. It is this possibility which is tied up with the notion of 'closed time'.

A concept of the first kind was put forward by Stannard (1966) on symmetry grounds. Like Wiener (1948), Grünbaum (1973) and Wheeler (1967), Stannard regards it as unlikely that there can be any means of communication between observers in temporally opposed systems. A signal 'departing' from the one system cannot be regarded as 'arriving' at the other, since that would be to deny the supposed opposite time directions of the two systems. The meaning of their putative 'co-existence' within the same space-time framework is therefore left extremely vague since co-existence depends on the applicability of a criterion of simultaneity and this is lacking if there are no means of signalling.

Type (2) 'reversed worlds' have been more widely speculated about and they involve two distinct mechanisms:

(2a) Reversal of the whole universe, or parts of it, by fluctuation;
(2b) Reversal of the whole universe by a quasi-deterministic process, as in the Friedmann model with $k = +1$, as already discussed.

It was, of course, the (2a) mechanism which was the subject of Boltzmann's speculations, very daring at their time.* As has been seen already, entropy fluctuations as small as 10^{-6} below the maximum within a laboratory sample of helium could not be expected to be observed in much less than $10^{10^{19}}$ years. Nevertheless if *infinite* time is available fluctuations of *any* magnitude might occur. In this sense Boltzmann (1895 and 1964) was well justified in regarding local parts of the universe as fluctuating during immense durations. "... there will occur here and there", he says (1964;446), "relatively small regions of the same size as our galaxy (we call them single worlds) which, during the relative short time of

* Boltzmann, as is well known, concerned himself with *thermodynamic* fluctuations. On the other hand E. P. Tryon (1973) has supposed that the universe appeared 'out of nothingness' by a *vacuum* fluctuation in the sense of quantum field theory.

eons, fluctuate noticeably from thermal equilibrium, ... For the universe, the two directions of time are indistinguishable, just as in space there is no up or down. ... In the entire universe, the aggregate of all individual worlds, there will however in fact occur processes going in opposite directions. But the beings who observe such processes will simply reckon time as proceeding from the less probable to the more probable states, and it will never be discovered whether they reckon time differently from us,..."

It's interesting that Boltzmann tacitly admits the unrefutability of his concept. This is much more than Reichenbach does writing sixty years later! Reichenbach (1956;128 ff) is concerned with 'upgrades' and 'downgrades' of entropy of *the whole* universe, and not just of parts of it; he arrives at the conclusion that "the universe consists of separate time threads pieced together in opposite directions ..." This, he says, is a meaningful statement "because time *order* can be defined in classical mechanics** and does not presuppose entropy. We may therefore speak of a *supertime* which orders the curve even in sections of equilibrium, where the entropy remains practically constant, or at saddle-points, where the entropy gradient reverses its direction. Supertime has no direction, only an order, whereas it contains individual sections that have a direction, though these directions alternate from section to section." Reichenbach proceeds to a detailed defence of this viewpoint, but nevertheless it troubles me. For it is implicit in his own account that observers within the universe could not refute his contention since *to them* the temporal direction is that of the 'thread' within which they exist. The opposed threads could only be known to some divine Observer standing outside the universe, and this gives a very metaphysical character to the whole theory. Nevertheless I fully agree with a much more modest, and indeed sceptical, comment made by Reichenbach on another page (1956;134): "The statistical nature of time direction appears to be the ultimate outcome of all inquiries into the nature of time."

Let's turn to the (2b) mechanism of reversal. As is well-known the question whether or not the universe is predicted to contract, at some time in the future, depends on the average density of matter and present evidence indicates that the density is probably not high enough. On the other hand 'hidden mass' may yet be discovered and may lead to the conclusion that contraction is to be expected. If so its onset may require little more than 10^{11} years – i.e. immensely less than would be expected for a major fluctuation to occur in the sense of mechanism (2a).

As has been seen, a contracting universe does not present any conceptual problem relating to its meaningfulness so long as processes such as radioactive decay, and the emission of starlight, can be expected to 'continue' unchanged and thus provide a temporal reference direction. On the other hand, if one goes further and postulates a continued sequence of expansions and contractions, the conceptual problem becomes immense

** He refers here to the 'betweenness' conferred by the causal theory of time as outlined in § 3.3.

since what may 'happen' at the singularities $R=0$ is quite unknown. Perhaps, as Wheeler speculated, 'the next' cycle will 'start off' as an entirely new universe and without any accumulated entropy, or vestiges of any sort, from 'the previous' cycle. He believes (1968) that space can have innumerable possible geometries and that at small enough distances there can occur a quantum fluctuation from one geometry to another. Although these fluctuations only become significant at distances of the order of the Planck length (10^{-33} cm) it is precisely as the singularity $R=0$ is approached that such fluctuations are to be expected. The concept of space-time, and with it 'time' as well, then breaks down. As Wheeler puts it, 'before' and 'after' lose all meaning and the notion of 'what happens next' becomes useless. If so, does not the concept of an oscillating universe become equally void of meaning? For how could supposedly 'successive' cycles be distinguished in the sense of 'the previous cycle', 'the present cycle' and 'the next cycle'?

Putting these difficulties aside let's consider a rather different one. Suppose it were a deduction from cosmological theory that the universe attains absolutely *identical states* during a sequence of oscillations. That this is not a red herring may be seen from a classic remark of G. N. Lewis (1930): "The precise present state of the universe", he wrote with an astonishing sense of certainty, "has occurred in the past and will recur in the future, and in each case within finite time."

This hypothesis leads to the question of 'closed time'. Everything in this chapter so far has posited that the temporal order is an *open* order, like the unending order of the real numbers. However this assumption seems to become untenable if the universe 'returns' to identical states – and of course if they are truly identical this must extend to the readings of clocks and calendars, and to the states of all memories and records.

In Grünbaum's discussion of 'closed time' (1973;197 ff) he keeps the linguistic pitfalls very much in mind. It is of crucial importance, he says, if pseudo-puzzles and contradictions are to be avoided, "that the term 'returning' and all of the preempted temporal language which we tend to use in describing a world whose time (in the large) is closed be divested of all of its tacit reference to an external serial super-time." Rather than thinking of some model universe 'returning' to an identical state at a different instant of open time, the universe would have to be conceived as 'returning' – in a highly Pickwickian sense of that term – "to the selfsame event at the same instant in closed time. This conclusion rests on Leibniz's thesis that if two states of the world have precisely the same attributes, then we are not confronted by distinct states at different times but merely by two different names for the same state at one time." *

* In Sklar's discussion (1974;314) he is inclined to reject this argument "since that version of the principle of identity of indiscernibles appears pretty implausible anyway." However, in Sklar's own discussion, especially of his Fig. 48, it seems that he makes entirely illicit use of a super-time in order to achieve his ordering t, $t+\Delta t$, $t=2\Delta t$, etc. The clocks and calendars which record these 'times' appear to be regarded by him as standing 'outside' the universe.

Grünbaum suggests that manlike beings could discover whether or not the time of their universe is closed; all they need do is to see whether the equations governing temporal evolution are deterministic, and whether the boundary conditions are such that all variables of state pertaining to everything in the universe (including those variables whose values characterize the thoughts of scientists living in it) "assume precisely the same values at what are prima facie the time t and a very much later time $t+T$. Then upon discovering this result by calculations, these scientists would have to conclude that the two different values of the time variable for which this sameness of state obtains do not denote two objectively distinct states but are only two different numerical names for what is identically the same state. In this way, they would discover that their universe is temporally closed, . . ." *

However this suggestion cannot lead to any present conclusion about time being closed since all existing theories relating to atomic phenomena are *in*deterministic. Not only macro-states but also all *micro*-states would have to be 'repeated' if the universe as a whole at time $t+T$ were to have precisely the same values of the variables of state as at time t. Thus there is no present evidence – in fact rather the reverse – that the universe is temporally closed. But suppose that quantum theory were to be overturned and that determinism were to be restored; there would still remain a very troubling feature of the concept of closed time and it is this: Let t in Grünbaum's discussion be the time and date, 11.14, Dec. 18, 1979, at which I am now writing, and let the predicted time, $t+T$, of an exact 'repeat' state of the universe, according to the hypothetical fully deterministic theory, be a time and date in, say, 12×10^{10} A.D. What happens as that time and date are 'approached'? At *exactly* $t+T$ the deterministic equations predict that clocks and calendars 'again' (using this word in the Pickwickian sense) read 11.14, Dec. 18, 1979. Will the equations predict the clocks an calendars as showing the time and date 'increasing' steadily right up to the very instant, $t+T$, in 12×10^{10} A.D. and 'then falling back' quite abruptly to the value 11.14, Dec. 18, 1979? Or alternatively will they predict the clock and calendar readings 'rising' to a maximum in some year short of 12×10^{10} A.D. and 'then gradually declining' to the present-day reading? It is difficult to conceive a deterministic theory, using the time variable t as we do use it, and regarding clocks and calendars as mechanical devices, which would predict *either* of these eventualities.

Moreover the words I have indicated in inverted commas ('increasing', 'falling back', etc.) tacitly assume that time, in some sense, is 'going on'. One has to ask: What is the criterion of this 'going on' if the behaviour of clocks and calendars, and all other physical systems, is in question? The conceptual problems associated with closed time are clearly immense.

---

* According to Tipler (1979) a performance of this calculation, using reasonable assumptions, does *not* result in the variables assuming 'precisely the same values', or even approaching arbitrarily close to them. This is because in general relativity, unlike classical mechanics, the singularities intervene to prevent recurrence.

Although Grünbaum's argument about what might be predicted deterministically at the instant $t+T$ seems reasonable enough, the result of the argument might be taken as providing yet another indication that determinism itself is unreasonable. Alternatively one might seek to exclude any consideration of closed time. For example, by postulating a sort of impotence principle to the effect that *not all* of the world's processes can be considered as undergoing reversals; one of them at least must be deemed not to do so. This would be in line with much that has been said already in this book.

§ 6. Other Asymmetries. The temporal asymmetries discussed so far, with the exception of causality, all have a strongly objective$_2$ character. Yet there are other asymmetries of a more subjective sort, and these will now be outlined.

Consider first our own sense of an irreversible temporal ongoing. As was said in Chapter 2, this is the primary basis of the ordering relationship 'later than'. Let's ask then whether it would ever be possible for us to experience the world's processes as being reversed relative to our own deeply ingrained sense of a temporal direction.

The answer will depend, I think, on where one makes the 'cut' between what is assumed to be reversed and what is not. Consider first a very distant cut – one which assumes a reversed motion of the distant galaxies, but of nothing nearer. It follows from what has been said already that it would probably be possible for humans to survive for many billions of years into a supposed contracting phase of the universe. Indeed the temperature of the background radiation might rise very considerably above 3K before the change even became noticeable. But let's consider the cut as being made very much closer in by supposing that solar and terrestrial processes were also experienced in reverse whilst our own bodily and psychic processes were not. Grünbaum and others have pointed out that we would experience very peculiar effects, such as motionless objects suddenly starting out into motion in a highly dangerous manner instead of coming safely to rest under the beneficent action of friction. Rain would be seen as rising rather than falling, rivers as flowing upwards into the hills, cars as moving backwards along the roads, and trees returning to seedlings, etc. The irreversible processes of the Second Law would seem to occur in the 'wrong' temporal sequence; bodies at an equal temperature might often become progressively different in temperature; gases might spontaneously unmix, and so on. Furthermore the external consequences of our acts of will would be experienced as occurring *before* the volitions themselves. All these peculiarities taken together seem to indicate that humans could not survive for long under such unfavourable conditions.

Let's now suppose that the cut were made still closer in, such that we experienced our bodily processes, as well as the external ones, as occurring in reverse relative to consciousness. We would find ourselves rising from the dead, becoming younger, and finally returning to the womb. Instead of eating food we would experience ourselves as regurgitating; our hands would take the food from our mouths and lay it on the table. Our eyes and ears would now be emitters of light and sound respectively, rather than absorbers. Our sense of touch too would presumably be quite ineffective since the nerve impulses would be travelling the reverse way along the fibres. In short it is unlikely that we could obtain any information at all about the external world.

On the other hand if the hypothetical 'cut' were eliminated, by supposing that consciousness is reversed as well, all would be expected to appear as normal! As J.J.C.Smart (1954) has remarked, if *all* processes were reversed, including the operation of memory, the world would turn out as being indistinguishable from our present world. Moreover there would then be no criterion for saying that anything *had been* 'reversed.'

The upshot of these remarks, I suggest, is that our experience of the sequence of events could not be otherwise than it is. Natural selection results in an adaptation to the one direction of the external sequence of events which makes life possible. Furthermore, since the living body depends on chemical processes, it may well be the case that our sense of 'earlier than' and 'later than', and also our rough estimates of duration, are conditioned by the entropic processes within our own bodies. * Or at least that is how an external Observer might see the matter, looking at us from the outside so to speak. Whereas to ourselves, as has been said already, our own awareness of the sequence of events has a logical priority over any of the asymmetries provided by science.

If we now accept: (a) that our innate sense of the temporal order does indeed correlate satisfactorily with the order provided by physical criteria, and (b) that we experience 'a present', interpreted as in Chapter 4 as a moment of heightened attention, certain other asymmetries can be understood.

First concerning why we have memories and traces of events which are earlier than the present but not of events later than the present. Recall from Chapter 4 that it is just as much accepted in the B-theory as in the A-theory that events *occur* at certain clock times. The two theories do not differ in that respect. To be sure the B-theorist holds that *all* events, whether consciousness judges them to be past, present or future, are equally 'real' in some sense. But this doesn't mean that he regards an event as occurring before the clock time at which it does occur. If a meteorite hits the Earth at 11.35 a.m. it *was* at a distance from the Earth's surface at clock times earlier than 11.35, and it *will be* at the Earth's surface at times after 11.35. The manifestation of its presence at these later times, either in the form of fragments or as a depression in the surface, is a trace or record of its arrival. Neither on the A- nor on the B-theory can there be any mechanism for this trace or record existing at times *earlier than* 11.35 since

---

* As is well known there are results from experimental psychology which support this conclusion; e.g. the fact that a subject counts what he believes to be seconds more rapidly if his body temperature is raised above normal, thereby increasing the rate of his metabolic reactions.

the meteorite was then at a distance. So also in regard to the action of memory. If I see a lightning flash at 5.0 p.m. it affects my sensory organs and brain at that moment; my central nervous system could not have been stimulated in that way before 5.0 p.m. because the appropriate photons had not arrived; but it can be affected in the form of a memory after 5.0 p.m.

So much for the reason why it is only those events which occur earlier than the present – *our* present – for which we can experience the existence of memories, or traces or records, during our *later* presents. As noted already, our awareness of the presentness of events (and this awareness is itself an event) is ordered consistently with physical criteria. It is a different question why the memories, or traces or records, *persist* – if indeed they do persist, for they need not. Yet there is nothing mysterious about their persisting. After all, when a door is closed it stays closed until some new event occurs; and similarly when a message arrives at the brain it stays in the brain until some new event, its possible deletion, occurs. We can, if we like, call the former event an *imprinting* on the brain, but this says little more than that an event has occurred and that its manifestations persist. And of course this persistence of the memory or record is future directed (and not past directed) for the same reason concerning the consistency of the temporal order which has already been discussed.

What has been said amounts to the asymmetry: we can have knowledge of events earlier than our present, but not of events later than our present. Our past is thus determinate and factual, whereas our future (if we are not determinists) appears as holding many possibilities. And for this reason again, we exert our wills toward the future and not toward the past.

Reichenbach (1956) sought to give an account of records directly in terms of entropy and more particularly in terms of 'orderliness'. He concentrated his inquiry on Schlick's well-known example of a footprint in the sand. He regarded the 'orderliness' of the footprint as having involved an entropy *decrease* in the sand, achieved at the expense of a greater entropy increase arising from the metabolic processes of the person who left the footprint. Records, said Reichenbach (1956;151), are states of frozen order. He also brought in information theory: the Second Law, he remarked (1956;167), "leads to the existence of records of the past, and records store information, . . ."

There are several things I take issue with in the Reichenbach account of records and traces. In the first place the identification of negative entropy with 'orderliness' is far from satisfactory unless the meaning of 'orderliness' is very carefully qualified (Denbigh 1971;55). This is a familiar point and need not be further pursued. Secondly, *a propos* of Reichenbach's remark that records "store information", it needs to be said that information theory, in the form developed by Shannon, is about a particular kind of 'information' and has nothing whatsoever to do with the *semantic* information which is conveyed by the footprint. Therefore it would be deceptive to say that the footprint *stores* information, in Shannon's sense, as one can correctly say that a magnetic tape is capable of storing a number of 'bits'. But it would be correct to say that the footprint *conveys* information in the semantic sense of the word.

My final point is that there are a great many traces or records which are not adequately characterised as being 'orderly' and which therefore provide counter-examples to the thesis that a record is a metastable orderly system surviving from the past. Think for instance of a window. It I find it intact I can regard it as a record of (a) having been inserted in the frame, and (b) *not* having been broken since it was last inserted. On the other hand if I find it broken I can regard it as a record of (a) having been inserted in the frame, and (c) having been fractured since it was last inserted. Thus in both instances the window is a record, and the relative 'orderliness' of the intact state and of the broken state is of no significance whatsoever in regard to the window's capacity to provide semantic information. The two instances differ simply in being records of different pasts.

A particularly striking counter-example is provided by Earman (1974) when he writes about coming across a bombed city: "... in what sense," he asks, "is the formation of the traces of the explosion (the ruined buildings, etc.) the formation of temporarily highly ordered states or the formation of subsystems of temporarily lower entropies than their surroundings?" Surely he is right; the bombed city is a record of bombing, but it displays neither 'orderliness' nor any clear indication of frozen-in states of lower entropy.

In my view most records and traces are not distinguished as such in any objective or physical sense; it is rather what *we* read into them that constitutes them as records or traces. Indeed there are very few physical objects which do not function in this way; if it is an oak tree it provides evidence that a sapling took root in this place, and furthermore that the particular tree survived and was not felled. If it is a lump of granite found in limestone country, it provides fairly reliable evidence that it was brought there by human action. And so on, and so on.

In short, a record or trace is an entity which recognisably persists as such (i.e. it is genidentical, although not necessarily changeless) during some temporal interval. If we come across it at a moment t during that interval we may be able to infer, from its state and location at t, and by use of our accumulated experience, which particular processes occurring earlier than t have contributed to its state. Therefore it is often erroneous to seek to understand records or traces in terms of orderliness. Furthermore their only necessary connection with entropy is that the temporal sequence of (a) their formation, and (b) the observing of them, is consistent with the universal temporal order whose objective$_2$ indicator is entropy increase; an event at a time t cannot have effects before it has occurred, but it may have effects which persist after it has occurred.

Yet another temporal asymmetry is that of prediction and retrodiction. In view of the foregoing it may seem paradoxical that there are situations where predictions about the future are far more reliable than retrodictions about the past. As has been said, we can *know* about the past, but not about the future. Evidently the circumstances which permit greater reliability of prediction than of retrodiction must be those in which, for one reason or another, we *don't know* full details about the past. We then have to use a probability judgment in either direction of time. (Indeed if we already *knew* what had happened there would be no sense in making a retrodiction!) The question to be asked is why the probability judgment towards the past is often less reliable than it is towards the future.

The reason for this can best be appreciated by considering systems which are fully symmetric in their circumstances in so far as they have been isolated for some time in the past, as well as remaining isolated for some time into the future.

As an example, suppose I find two blocks of metal at an *equal* temperature (so far as the precision of measurement allows). So long as the blocks remain isolated I can predict with great reliability that they will remain at an equal temperature. On the other hand as regards retrodiction I am helpless! I cannot say whether the blocks had been at an equal temperature ever since isolation, or whether they had been at unequal temperatures. Still less can I retrodict which of them had been the hotter.

This unequal reliability of retrodiction and prediction applies very widely. If I find two gases well mixed I can predict that they will remain well mixed, but I cannot retrodict that they were so in the past. Similarly if I find the chemical reaction $A+B=C+D$ in an equilibrium state I can predict that, in the absence of external influence, it will continue in that state, but I cannot retrodict that a little earlier the mixture might have had a non-equilibrium composition. Nor can I say from what *particular* non-equilibrium state an observed equilibrium state has been obtained.

From the thermodynamic viewpoint the matter may be summarised as follows: Using the Second Law it can be reliably predicted that isolated systems will approach equilibrium in the future; or, if they are already in equilibrium, it can be reliably predicted that they will remain so. On the other hand science provides no corresponding means of retrodicting how these systems might have arisen, in their given states, from the past; they might have arisen *in an immense variety of ways* and therefore, if their past is not actually *known* from observation, or if it is not inferable from records or traces, it is not retrodictable. Prediction within isolated systems is towards more probable states of greater uniformity; retrodiction, on the other hand, would have to be towards *less* probable states of lower uniformity. Unless reliable records or traces are available, the backwards extrapolation towards the *particular* special state in the past, from which the presently observed state has arisen, cannot be carried out.

The matter can also be discussed from the viewpoint of Bayes' theorem in probability theory. Willard Gibbs (1902) made the essential point very clearly: "... it is rarely the case that probabilities of prior events can be determined from those of subsequent events, for we are rarely justified in excluding the consideration of the antecedent probability of the prior events." This theme has subsequently been developed, both in a classical and in a quantal manner, by Watanabe (1955, 1966, 1969) and Costa de Beauregard (1963, 1965, 1972). The problem of retrodiction, says Watanabe, is that of an observer B who wishes to guess from his own experimental data the result which another observer A earlier obtained on the particular system. The difficulty for B is that he has to assume an a priori probability for each possible initial state in which A might have found the system. In general these antecedent probabilities are unknown and there can be no good reason for taking them to be equal, for each quantum state; except, he goes on to say, when the initial ensemble given to A can be assumed to be the result of an ergodic process. (See also d'Espagnat, 1976.)

§ 7. Conclusion. As has been seen, the three concepts of 'time' which arise from theoretical physics, from thermodynamics and from conscious awareness respectively differ from each other most conspicuously in regard to the amount of reality they attribute to 'the present' and to 'time's arrow'. In Chapter 4 it was shown that very little evidence can be adduced, one way or the other, on the question of the objective reality of 'the present'. In the light of the present chapter what can now be said about the other question, the reality of the 'arrow'?

Throughout the chapter continued expression has been given to the view that 'time' is not some sort of existent which has a reality apart from events and processes. We can, if we wish, speak of 'time's arrow' or of 'the direction of time', but what we are referring to is much better regarded as an asymmetry or anisotropy of the *temporal order,* as indicated by certain kinds of process. As such it may perhaps not apply universally or to all theoretical epochs. It has been shown in fact that there is no *absolute* criterion of temporal anisotropy; the 'arrow' provided by thermodynamics is probabilistic, and so also is the 'dispersion arrow' – the spreading outwards from a centre of waves or particles. As for the criterion of anisotropy which is supposedly offered by the expansion of the universe, there is at least one respectable cosmological model which allows of re-contraction and this would not be a self-consistent notion unless one of the other 'arrows' were taken as providing a reference direction. For it is meaningless to think of 'increasing' values of the time variable t unless there are *at least two* distinguishable processes one of which is deemed not to undergo any 'reversal' and which therefore provides a reference process.

Furthermore none of these physical criteria *directly* display events as being earlier or later than each other. What we do find in nature is a certain *parallelism* of processes such as is expressed in Schrödinger's formulation of the Second Law and in the Reichenbach-Grünbaum theory of branch systems. And of course this parallelism of processes displays a highly reliable correlation with the human sense of earlier than and later than, but the correlation remains empirical and contingent.

Relative to the human sense it cannot be denied that gases tend to mix, that bodies in contact become more nearly equal in temperature and that nuclear processes, such as the conversion of hydrogen into helium in the stars, continue steadily for countless billions of years. All such processes provide 'arrows' which are consistent with each other and they can be made understandable on the basis that the universe is evolving from an 'initial' state which contained an immense potential for further change. The empirical evidence for temporal anisotropy, during the present epoch and in spite of the *t*-invariance of the basic theories, is thus extremely strong. If we require a generally reliable and objective$_2$ criterion of the order of events, a criterion which would be applicable in man's absence from the world, we should clearly look to just those entropy-creating processes, such as the decay of radioactive isotopes, which we actually use in the dating of events in the remote past.

Yet it would be a grave mistake to regard any one of these processes as providing a *definition* of 'time's direction', since any particular system, or even a large ensemble of systems, could conceivably undergo reversal due to fluctuation phenomena. It is far better to regard 'the arrow' as being an *emergent* feature of those physico-chemical systems which are large enough for fluctuations to become exceedingly infrequent. As Reichenbach has said, 'time's direction' has a statistical character. Thus none of the physical criteria, with the possible exception of the dispersion of particles or photons into a supposedly infinite universe, can provide certainty.

Moreover our understanding of how the universe may have 'begun', or of what may 'happen' at the $R=0$ singularities (and indeed whether science as we know it is applicable to such extreme states), is so nebulous that it is unreal to conceive of a temporal direction which would be valid 'for always'. Temporal concepts, however essential they may be for dealing with experienced phenomena, should not be stretched too far and in particular the notion of 'for always' seems meaningless.

Let me revert briefly to the *A*- and *B*-theories where the latter asserts that all events along a world-line are equally real whereas the former predicates the reality of 'the present' and of 'becoming'.

In this chapter it will have been seen once again how effectively the tacit use of the *B*-theory assists in the thinking out of the various issues concerning time. It provides a model of events as if they were beads along a string. But of course this facilitation of thought does not establish the *B*-theory as being true! Furthermore its static character seems to present us with a very peculiar asymmetry: low entropy in the one direction along the world-lines and high entropy in the other direction. As was said earlier, this asymmetry appears much more readily understandable if it is supposed, as in the *A*-theory, that there is a real process of 'becoming'. Although science offers nothing which can provide a decision between the two theories, which therefore appear as unrefutable, it will be seen in the next two chapters that the *A*-theory provides the more attractive metaphysical basis for the life sciences just as the *B*-theory provides the correspondingly more attractive basis for physics.

Chapter 7

Temporal Ongoings in Biology

§ 1. Introduction. As was said in Chapter 1, it was formerly 'read into' physical science that the universe is passive and inert, and has no initiating powers of its own. This was due to the influence of determinism, and indeed the possibility of discovering scientific laws had seemed to depend, to many of the classical physicists and also to many philosophers, on the reality of a strict determinism in the occurrence of events. The very existence of physics and chemistry appeared to require that everything takes place as if laid down, *in posse,* at The Beginning.

Biology, by contrast, has envisaged a world in which radically new things come into being 'in the course of time', and this is a view which has become much more acceptable to modern physics subsequent to the development of quantum theory. The notion of scientific 'laws' can now be understood either statistically, as applying to an ensemble, or probabilistically as applying to the uncertain behaviour of any individual. And of course this change of outlook has not been accompanied by any loss of predictive power in experimental situations; indeed quantum theory has been immensely successful in this respect.

In an earlier book (1975) I argued that this new-found openness of outlook within science allows of natural processes being regarded as having an inventive or creative character. By this I mean nothing anthropomorphic – except, of course, in so far as all interpretation is inevitably coloured by the ordinary usage of words. I mean no more than that natural processes are capable of bringing genuinely new things into existence – things, that is to say, which are not physically or logically necessitated by other things existing previously. *

To be sure it does need to be assumed – as I shall do throughout this chapter – that there is something real about the humanly experienced ongoing of events, as is implicit in the words of the previous paragraph. This is tacitly to adopt some of the features of the A-theory of time, and, as has been seen in Chapter 4, there is as good a case for this theory as for its

---

* Elsasser (1969) has a rather different concept of natural creativity as meaning the capability of a random substrate to give rise to ordered patterns by indeterministic processes.

rival. Thus when I speak of new things coming into existance, I mean in the time direction of consciousness, here regarded as being *the* direction.

No doubt the notion of inventiveness in nature is just as much a gloss as was the passive picture. It may add to our total comprehension of the world we live in, but can hardly be expected to result in any falsifiable propositions. **

I use it only in opposition to that impression of the deadness and inertia of the natural order which was conveyed by classical physics. To speak of an inventive universe is no more than to make two conjectures: first that the universe is indeed an evolving system, and secondly that the course of its evolution is such that entirely novel entities are occasionally produced as the result of chance events.

It is of little use to look to cosmology for any support. This is such an unsettled and speculative subject as to be of dubious assistance one way or the other. To be sure it is very generally agreed that the universe is in process of change, and furthermore that important aspects of its evolution may have arisen indeterministically – for example because of fluctuations or of various forms of instability. Yet this remains hypothetical and uncertain. In the following I shall therefore refer no more to cosmology but instead develop the notion of inventiveness in the area where it can be most adequately supported – i.e. within the life sciences.

§ 2. Reconciliation with Reductionism. The merits and demerits of reductionism have been widely discussed and here I will add only one brief point of my own.

In their well known early paper on the subject, Oppenheim and Putnam (1958) regarded the 'elementary particles' as the most basic level and thus all 'higher' levels should eventually be capable of being 'reduced' to the properties of these particles. But what if no ultimately basic level exists? Mention has been made already of Bohm's views of 1957 and these have obtained some support from the results of subsequent research. The particles which only a few years ago were supposed to be 'elementary' are now assumed to be made up of others, such as the quarks and gluons. But already this scheme has lost its pristine simplicity since it has now become necessary to postulate several kinds of quarks. If, as Bohm supposes, there is an infinite regress of particles the scientist can never achieve a finite reductive scheme. Moreover this is not an issue which conveniently confines itself to the mysteries of nuclear structure since what happens (unpredictably) in the nucleus can have consequences at all higher levels. For instance, as was said in § 5.3, the energy-yielding reactions between particles greatly affects the theory of stellar evolution.

In spite of this caveat – and there are others which might be put forward such as the failure adequately to 'reduce' thermodynamics – there can be little doubt that 'reductionism', when this term is used in a less

** In a later section a testable conjecture concerning the effect of the incidence of chance events on organisms will nevertheless be put forward.

strong sense than it is by the theoreticians of the subject, plays a valuable part in scientific methodology. The scientist does indeed try to understand the properties of complex entities in terms of the properties of the simpler entities which they contain. This applies with particular force to those scientists, such as biochemists and biophysicists,who work at an interface. The value of the contribution made by DNA studies to the linking of biochemistry with genetics, cytology and microbiology can hardly be overestimated.

Popper, in his comprehensive review (1974a), concludes that the successes of reductionism have always been partial and have invariably been accompanied by unexplained residues. Yet these residues have been immensely helpful in stimulating further research and for this reason the faith in reductionism, like the faith in causality, remains invaluable to the practisting scientist. "As a philosophy", he says, "reductionism is a failure. From the point of view of method, the attempts at detailed reductions have led to one staggering success after another, and its failures have also been most fruitful for science."

Is it the case then that 'nothing buttery' is still unproven? The issue has been much discussed (e.g. Koestler and Smythies, 1969; Grene, 1971; Lewis, 1974). One school of thought points to the philosophical inadequacy of reductionism and to the necessity, even in science, of using higher level concepts such as truth and consistency, a matter which will be referred to in Chapter 8. The opposing school gives emphasis to the practical successes and maintains that an anti-materialist philosophy cannot safely rely on the unexplained residues: "It's chemistry, brother, chemistry!", says Mitya in *The Brothers Karamazov,* "There's no help for it, your reverence, you must make way for chemistry."

My own leaning, like that of most scientists, is towards naturalism – and this I distinguish from materialism and from physical monism in that naturalism accepts that sentience and consciousness are realities within the natural order. When it is so regarded, nature is immensely richer and more diverse than the older forms of materialism would suggest. Arguments from continuity serve to indicate that sentience is latent, in some sense, at the very roots of the reductionist scheme – that is to say as a *possibility,* and one which has in fact been realized because the appropriate inventive processes have occurred.* Such a view is to enrich, and not to impoverish, the general conception of nature.

Let's move on to less well-trodden ground by enquiring whether the inventiveness I have postulated can be reconciled with reductionism.** But first I must amplify what was said in § 1.

* These remarks are not intended as advocating panpsychism. A skillful and scholarly application of Whitehead's panpsychism to biology has been made by W. E. Agar (1943), but I am not convinced that the attribution of a primitive kind of sentience to all cells within the living body is really distinct from the supposition that the cells are programmed to perform their function by the genetic apparatus and by the effect of the morphogenetic field.

** What follows is an expanded version of material I had already presented in *An Inventive Universe* (1975).

A process may be described as inventive if two conditions are satisfied: (1) the process produces something which is qualitatively different from anything existing previously; (2) the occurrence of the process is not physically or logically necessitated. The conjunction of these independent conditions provides the meaning of what has so far been referred to rather loosely as 'genuinely new'. Even so some further amplification is needed.

The question whether some particular entity is qualitatively different from all previous existents clearly raises some important issues about criteria, but it would be out of place to go into these in detail. Sufficient to say that a decision concerning difference can often be based on an act of appraisal such as is made by the taxonomist. In other instances however it may require a very detailed comparison of physico-chemical properties. The failure of the Viking expedition of 1976 to establish whether or not there are organisms on Mars is a case in point.

Condition (2) above raises much deeper philosophical issues but these have already been considered in Chapter 5. Let's just recall that predictive determinism (but not the ontological doctrine) was seen to be a reasonable thesis and moreover it allows of the concomitance of chance and causality. In many instances the occurrence of an event or process is quite unpredictable precisely because the relationships or patterns which are actualised, which become real, are affected by extraneous factors within the past light cone. Consider the example of the DNA-protein system. Rather than supposing that it is *necessarily* what it is, a more reasonable hypothesis is that it represents but one possibility out of a vast number of alternatives, any one of which *might have been actualised.* In other words our scientific understanding of this system is enhanced by regarding it, not as having been necessitated, but rather as being the product of innumerable chance factors.

My notion of inventiveness in nature is clearly somewhat similar to the concept of 'emergence'. Yet they are by no means the same. Emergence, as it is usually understood, has been rightly criticised on the ground that it involves discontinuities in the form of new principles or laws suddenly becoming operative at various levels. If so it is contrary to reductionism. Thus according to Schlesinger (1963;48) the thesis of emergence means that "when objects of a given level combine to form wholes belonging to a higher level, the properties that emerge are not reducible to the properties the elements possess when separated." On the other hand Popper (1972;298) has written: "... unpredictability in principle has always been considered as the salient point of emergence ..." This is close to my own concept and I shall proceed to show that although inventiveness in nature depends on unpredictability it is in no way contrary to reductionism.

Oppenheim and Putnam proposed the following 'levels' as being natural and justifiable from the reductionist viewpoint: elementary particles, atoms, molecules, cells, multicellular organisms, social groups. Let's consider a more general n-level scheme: $L_1, \ldots L_i, \ldots L_n$. The reduction of, say, L_i to L_{i-1} requires the existence of a theory, call it T_{i-1}, which is capable of accounting, in terms of the entities existing at L_{i-1}, not only for all of the phenomena of L_{i-1} but also for all of the phenomena of L_i. Let it be supposed that the n-level scheme has been chosen as *an evolutionary sequence* as was indeed the case with Oppenheim and Putnam's choice. (The elementary particles came before the atoms, the atoms before the molecules, etc.). There will thus be a time t_{i-1} when the level L_{i-1} first appeared and there will be a later time t_i when the higher level L_i first appeared:

$$\text{At } t_{i-1} \text{ the levels } L_1, L_2, \ldots L_{i-1} \text{ exist}$$
$$\text{At } t_i \text{ the levels } L_1, L_2, \ldots L_{i-1}, L_i \text{ exist.}$$

We ask whether the process $L_{i-1} \rightarrow L_i$ can be regarded as inventive and this will be the case if the formation of L_i was not necessitated. It is just this possibility which is opened up if ontological deteminism is denied. For in that case the entities of L_i may arise by chance; perhaps through the occurrence of appropriate thermodynamic fluctuations or through the arrival of suitable activating particles from outer space. Thus at all times prior to t_i the theory T_{i-1}, however good it may be, is incapable of asserting that L_i *must* appear and the process then qualifies as being inventive.

To be sure at times prior to t_i there may be several competing theories. One of them, call it T'_{i-1}, may indicate that L_i *might* appear, along with other possibilities. Another theory T''_{i-1} perhaps fails to predict the formation of any higher levels whatsoever. But unless there is a theory prior to t_i, which shows that the appearance of L_i is necessitated, then the appearance of L_i, when it does occur, is the appearance of something which is genuinely new.

Another important point is this: as soon as L_i has come into being a reconsideration of the existing theories is required. Take the case of T'_{i-1}; if this had preticted L_i as being one among a number of other possibilities it might still be regarded as a satisfactory reducing theory for L_i. (A case in point is biochemical theory; this is generally regarded as a fairly satisfactory reducing theory for the understanding of many cellular processes but it cannot assert that a genetic code *had to* come into being or that the particular code which exists is the only one which was possible.) On the other hand if there were no theories appropriate to the level L_{i-1} which could predict the formation of L_i, even as just a possibility, the foundations of the existing theories would need to be overhauled. This might involve deep conceptual changes or it might require little more than the attribution of a previously unknown property to the entities existing at level L_{i-1}. An example was the attribution of *spin* to electrons. This was done later than the original formulation of quantum theory and was required in order to enlarge the scope of the theory's reducing powers. The Pauli Exclusion Principle was also introduced into the theory for similar reasons.

In short it has been seen that the ascription of an inventive character to the natural order requires only the denial of rigid determinism and does not require that 'vital forces', or the like, suddenly become operative at various levels. Thus no conflict with reductionism is entailed.

That the foregoing is consistent with present day scientific theory may be seen by referring back to Oppenheim and Putnam's six-level scheme.

Currently we are at a time t_6 when 'social groups' provide the highest existing level, and we have no theory whatsoever which is capable of asserting that some yet higher level, L_7, will eventually make its appearance. Thus, if *per mirabile* a level L_7 were to come into existence, there are no grounds on which it could be said to have been necessitated; its formation would be the result of an inventive process in nature.

§ 3. Chance and the Organism. There are comparatively few natural varieties of gases or liquids, or even of minerals, but a truly immense variety of organisms. For instance our own phylum of the Chordata contains some 40,000 described species and the phylum of the Arthropoda contains over 900,000 (Hanson, 1964). No doubt a similar diversity has occurred on other planets elsewhere in the universe which have been graced by the touch of life – their biospheres have quickly blossomed with a vast profusion of living forms.

This inventiveness of living material is made clearer still by the fact that, among those species which reproduce sexually, each single organism is very probably a unique individual. The number of heterozygous genes is often large enough for the number of possible combinations of those genes vastly to exceed the number of members of the particular species (Simpson, 1950; Dobzhansky, 1963; Elsasser, 1975). As a consequence each separate member of the species (apart from the case of identical twins where there has occurred a division of the early embryo) very probably has a unique genotype. In short the process of sexual reproduction is *continuously inventive.*

The sexual process also illustrates very nicely the interplay of chance and causality. It can certainly be said that the fertilization of the egg cell is 'caused' – it is caused by the penetration of the egg cell by a sperm and by the subsequent fusion of their haploid nuclei. Yet it is also a chance process. As is well-known enormous numbers of sperm, some 10^8 in man, are released by the male and they propel themselves about, within the female genital tract, until one of them collides with, and penetrates, the ovum which then develops a resistance to any further penetration. How could it ever be predicted that *a particular* sperm, out of the many millions which are present in the tract, will succeed in fertilizing the ovum? Even if the movements of the sperm could be predicted, through some fantastic calculating feat on the part of a hydrodynamicist, it could not be known that the winning sperm has some particular genotype. Yet the genetic make-up of the embryo which results from the union depends on the genotype of the successful sperm, as well as on that of the ovum.

In the foregoing there is no inventive *agent* – unless it can be said to be 'nature itself'. On the other hand in the case of man, and perhaps some of the other primates to a much lesser degree, each individual is an inventive agent in his own right. This first shows itself in childhood in activities such as the creation of new forms of play. And also of course in everyday speech – in the putting together of words to form sentences having a detailed structure such that there is a fairly high probability that they may never have been uttered previously in precisely that form. It shows itself most conspicuously in the immense ceative output of a Shakespeare or a Mozart, a Newton or an Einstein.

What is it, we may ask, about the living creature which has enabled it to attain this prodigious acceleration of inventiveness? The conjecture I shall put forward in answer to this question is that living creatures are so organized that they can not only cope with chance factors in their environment but more importantly *they can profit by them.*

The usefulness of chance events to organisms can perhaps be summarized as follows:

(a) To the species: variation and hence adaptability
(b) To the individual organism: stimulus and the coming of new opportunities
(c) To the human: a range of possible futures; mental creativeness.

Here I shall deal only with (a) and (b), deferring some remarks about (c) to the following chapter.

Let me begin with some familiar material which is included only for completeness. As is well known, mutation is not the only source of variation in the genotype. Another is cross-over among the chromosomes and yet a third is the randomisation which occurs in sexual reproduction. There are no present indications that individual occurrences of variation will ever be predictable. In the case of mutation by radiation, for instance, it would require that it be knowable that a cosmic ray particle of sufficient energy will arrive at a particular instant and will strike a particular DNA molecule at exactly the right spot. The biologist deals with *populations* of organisms and thereby he establishes the *probabilities* of mutations – e.g. that they occur once in every 5000 generations of a particular strain under specified laboratory conditions. The apparent precision of such a result is in no way contrary to the chance character of the individual events.

Monod, in his provocatively entitled book (1972), accepts the chance character of variation but seems to regard the subsequent processes of selection and amplification as occurring of necessity. However this is surely to make too rigid a distinction. These processes are not completely error-free; selection and amplification are still subject to chance events although much less frequently. It is simply that the probabilities of certain outcomes approach much closer to unity. *

Biologists and biochemists have shown that selection and amplification take place in distinct stages. That is to say there are chemically selective processes within the individual cells and these precede the Darwinian processes due to the environment. The former include the template

* Waddington (1974) makes a similar comment on Monod. On the other hand Jacob seems at first sight to take a Monod-like view for he writes (1974;2): "In the chromosomes received from its parents, each egg therefore contains its entire future: . . .". But elsewhere in his book he greatly modifies this notion of a rigid future for the organism.

processes which occur in DNA replication, the coding processes by which RNA creates the proteins, and thirdly the highly specific catalytic actions of the enzymes. For these complex molecules are folded over on themselves so precisely that only a single kind of simple molecule, out of the many thousands of kinds available in the cytoplasm, can fit into the catalytic site within the enzyme structure.

As has been said, these chemical selection processes are followed, if they have been error-free or if they have produced a viable variation, by a process of amplification. This is not yet well understood but it appears to involve programmed sequences of reactions such that, accompanying any one of them, there are signals which specify that another must follow (Stebbins, 1969). Amplification is essentially a continuance and a build-up of what has already been achieved by selection. Indeed Eigen (1971) has shown that the occurrence of autocatalytic and non-linear amplification is equivalent to very sharp selection – so sharp that one, and only one, development wins out at this biochemical level.

So much very briefly concerning chemical selection and amplification. These processes are fairly error-free and therefore are potent factors in ensuring a continuation of the normal life of the organism and a high degree of stability of inheritance. During the earliest historic stage of evolution, the supposed period of the 'biochemical soup', these chemical processes presumably provided the only means for selection and amplification. But of course during the later stages Darwinian selection and amplification also came into prominence.

Darwinian selection itself is so familiar in outline (although still highly controversial in detail) as hardly to require any description. Put very briefly, those genetic variants of a species which adapt themselves most successfully to a particular environment, undergo a second stage amplification which amounts to an increase of their population relative to the less adaptable variants. What originated as a chance event at the molecular level thus finally becomes stabilised and results in a new way of life. Jacob (1974;296) expressed it well: "Evolution is built on accidents, on chance events, on errors. The very thing that would lead an inert system to destruction becomes a source of novelty and complexity in a living system. An accident can be transformed into an innovation, an error into success."

What is it then that enables a living system, as distinct from an inanimate system, to achieve this? It has been suggested already that it is the quality of 'being organised'. If one spoke teleologically, one might say that the living creature is organised *so as to achieve* what it does achieve. Yet there is no necessity for this kind of implication; it can simply be said that an organism's particular state of organisation is such that it results in what it does result.

The question concerning what is meant by 'organised' will be deferred to the following section. For the moment I shall deal only with my conjecture about chance events. It has been seen from the foregoing brief survey that the appropriate model for the inventive process, as it occurs in living things, is one which begins with chance events, such as mutations, and these are followed by selection and amplification stages. This results in the adaptability, not only of the whole species, but also of its individual members when they meet with new situations. Chance is the origin of opportunities. Does it not seem likely therefore that one aspect of Darwinian selection has been the selection of those particular organisms which can best cope with chance events and can best seize hold of the opportunities which chance events offer?

Notice that there is an important distinction to be drawn between the *coping* with contingency and the *utilising* of it. This suggests that for each species, and also for each individual within that species, there may well be an optimum intensity of chance effects, or an optimum frequency of their occurrence. Below and up to the optimum point, the species or the individual will be able to achieve benefit; but if the optimum is exceeded the organism's self-regulative system will be overwhelmed and unable to cope.

This hypothesis obtains a measure of support from certain well-known phenomena. Think for instance of the results of experiments on sensory deprivation. When there is an almost complete absence of accidental changes in the environment, the organism suffers through lack of sufficient stimulus. On the other hand, excessive environmental changes are also harmful. A living creature cannot tolerate extreme alterations of, say, temperature or oxygen partial pressure. Between the conditions of sensory deprivation, at the one extreme, and those of excessive environmental variation, at the other, there clearly exists an optimum.

Some further support is provided by genetics. Let us ask why it is that organisms which reproduce sexually cannot, in general, fertilize their gametes except with members of their own species. On the face of it, interspecies mating might appear advantageous since it would give rise to even more variation than is obtained by normal mating. Why then has it not been favoured by natural selection? Although I haven't found this question discussed in biology, it seems likely that interspecies mating would give rise to an altogether excessive degree of mismatching. The resultant gene complex in the zygote would be such a mixed bag as to impair the proper biochemical functioning of the cell. Even within the same species, polyploidic strains are usually quite infertile with each other; presumably their matings would not result in a sufficient degree of matching of the chromosomes. The frequently deleterious effect in man of even a single chromosome mismatching is well-known.

Yet another argument to the same effect is provided by the sameness of the genetic code in all terrestrial organisms which have been examined so far *. It has been suggested that the code was formed at a very early stage of evolution, and has subsequently remained unchanged during a vast period which has seen the appearance of all the various phenotypes. If this

* Recent evidence (Hall, 1979) has shown however that the mitochondria in various species use a genetic code which is independent of, and distinct from, that which is used in other organelles within the cell.

is correct, how is it to be explained? Surely on grounds of stability. Let's suppose that some particular codon, wherever it occurred along the length of a DNA molecule, in some particular cell, had started to code for a *different* amino acid. The consequence, as Luria (1973;52) has remarked, would have been a change in the structure of *almost every protein* produced by that cell. A change of this magnitude may well have been far greater than the organism could tolerate; it would have exceeded by far my postulated optimum of chance effects. Thus even though such 'mistakes' may indeed have occurred, they have presumably been eliminated through death of the particular cell. This provides a reasonable explanation why all species, ranging from viruses to man, have been found to use the same genetic code. No doubt a similar argument could be adduced to account for the fact that all species use the same repertoire, meiosis and mitosis, of cell division.

Having this amount of support, I am encouraged to put forward my conjecture rather more positively in the form of a diagram. Let it be supposed that biologists have attained some method of measuring the advantage of chance effects, either to a species or to one of its individual members. Imagine the measure to be plotted three-dimensionally against the intensity and frequency of chance events, as these are encounted by the organism. The result, according to the conjecture, would be somewhat as shown in the figure.

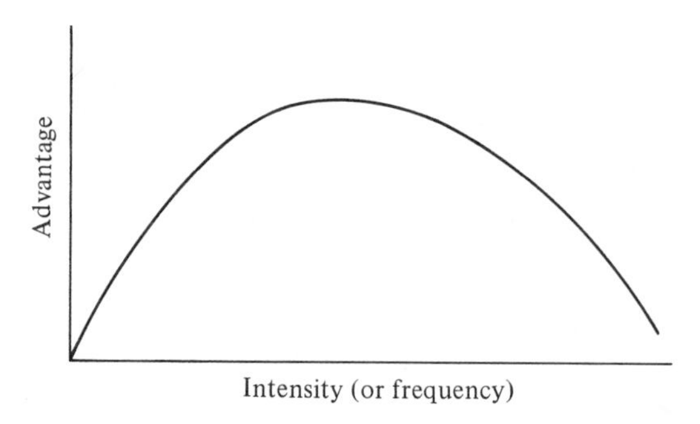

Of course the hypothesis embodied schematically in this figure is distinctly speculative. Nevertheless it is far from being metaphysical since it is capable of being tested – e.g. by controlled psychological experimentation involving random events of varying intensity and frequency.

No doubt the position of the optimum would be found to vary widely from one species to another, and also, although less markedly, from one individual to another. Many simple organisms do not have a regulative system of sufficient complexity to allow them to adjust to severe external fluctuations. And again, within a particular species, there may be appreciable variations in the ability of individuals to cope with the unexpected. Among humans, for instance, the young and the highly creative tend to thrive on what chance throws up. The very old, on the other hand, require a considerable degree of regularity in the environment.

Even so, it would seem that man is exceptionally well endowed, as a species, both to cope with chance and to utilise it. Due to his powers of imagination and anticipation, together with his memory, he can bring a very powerful selective system to bear on his search for opportunities. Relative to other species, man's optimum may be expected to be far to the right in the diagram. His organisation is such that he can adopt flexible strategies.

§ 4. What is Meant by 'Organized'? For the purposes of § 5 I need to attempt an analysis of this vague but useful word 'organized' which has been prominent already. As a first step it may be noted that it is used in both a temporal and a non-temporal sense. For instance in the case of an organism one can speak of the temporal organization of its development, its life cycle and momentary behaviour. One can also speak of its 'state of organization' and this refers to some quality it possesses at an instant, so to say.

A truly marvellous example of temporal organization is the whole process of morphogenesis. A single fertilized egg cell divides repeatedly and gives rise eventually to millions upon millions of daughter cells – some 10^{14} to make up a human body. In vertebrates the cells differentiate initially into three tissue layers – ectoderm, mesoderm and endoderm – and these subsequently differentiate further and give rise to the various organs and tissues. The cells which go to form the liver are all made in the appropriate place, those which go to form the eyes are made in other appropriate places, and so on. All these processes are accurately timed so that the various organs and tissues are created at mutually related rates, resulting for instance in the limbs growing at a suitable speed in relation to the rest of the body. What is particularly remarkable is that, although all the somatic cells of the particular organism have the same genetic make-up, signals or chemical gradients are nevertheless available at the various locations and these result in particular genes being switched on or off as required for the formation of liver cells, cornea cells, etc. Thus at each stage of morphogenesis there is a 'state' of organization which may be said to control development at that stage, but the 'state' is continually changing, at least up to adulthood. And of course even during maturity new cells are being formed to replace old ones. The body is an open system, in the thermodynamic sense, and most of its parts are being continually renewed.

Another aspect of temporal organization concerns the life cycle, such as is shown most remarkably by migratory animals like the salmon and the arctic tern. And yet a further aspect is the momentary behavior of the organism in relation to environmental changes. By means of multiple feedback loops, and other forms of cybernetic control, many organisms can achieve a high degree of homeostasis in regard to their own internal states and, at the same time, make appropriate adaptive responses, such as the seizing of food, to whatever the environment may suddenly provide. Indeed at the level of some of the arthropods and of many of the vertebrates, it becomes appropriate to speak of trial-and-error learning and of problem solving.

The foregoing describes the organism in *functional* terms – i.e. in terms of *what it can do.* * The weaknesses of functional description have been carefully analysed by Nagel (1961a), Hempel (1965), Lehman (1965) and Wimsatt (1972). As the latter has put it: "... the consequences of an entity which are chosen as its functions depend upon the perspective from which that entity and its consequences are viewed." In general there is no isomorphism between structure and function; there can be many alternative 'parts' for performing the same function, and therefore it cannot be said that some particular 'part' is essential – i.e. that there are no alternatives. But equally it cannot be said that there are not alternative 'functions' which adequately provide for survival. As Wimsatt says, the functions depend on the perspective. Take for instance the circulation of the blood; in a certain sense the heart may be said to be 'necessary' for pumping – yet a heart is *not* necessary to every organism since there are many which do not use blood and have no heart.**

For these reasons many biologists regard it as more scientifically profitable to give their attention to 'states of organisation' – i.e. to the consideration of structural arrangements and interconnections regarded as a momentary or enduring state. But what does a 'state of organisation' mean? This is the essential question. For obviously not *any sort* of structural arrangement of 'parts' will count. Does this entail therefore that only those particular arrangements of parts which are capable of fulfilling a useful function will count as being organised? If so we seem to be involved in a vicious circle.

Notice as a first step that the concepts of 'organization' and 'order' are by no means the same. Or to put it more accurately their meanings seem to diverge. In their applications to lower level entities the terms 'organized' and 'orderly' might be used interchangeably, as if they were synonyms. The structure of an atom or molecule might be described equally well in either way. Yet at a higher level the terms are by no means interchangeable and indeed their erroneous conflation by certain authors has led to serious mistakes. It has been widely supposed, for instance, that when an organism becomes more highly organized, during morphogenesis, this must necessarily imply a reduction of entropy.

The error may be clearly seen from the existence of pairs of entities such that one member of the pair must be regarded as *more* highly organized but *less* orderly than the other. Such a pair is a living cell and a crystal. Even though we have only an intuitive understanding of 'organized' we can hardly fail to regard the cell as the more highly organized member

* Of course the 'functions' do not offer any unique *definition* of life since they can be simulated by artefacts, although no doubt much less efficiently.

** Lehman (1965) points out that there must be hearts in some organisms before we can make a function statement about hearts. Thus the knowledge of the existence of the particular organ *precedes* any statement about its function and therefore this statement cannot explain the occurrence of the organ.

of the pair. Yet it is less orderly; it displays no accurately reproducible internal pattern and no accurate repeating distances such as are found in a crystal lattice. Any single example such as this which shows a reverse order of ranking is sufficient to show that the attributes in question cannot be the same. ** Thus even if 'orderliness' can be related to negative entropy there is no reason to suppose that degree of organization can also be so related.

The usefulness of the term 'orderliness' diminishes as the series of 'levels' is ascended, due to the appearance of greater individuality among entities of the same class, but the significance of the quality 'organized' simultaneously *increases*. The latter, in my view, is a distinctively 'higher level' concept and we obtain our primary understanding of its meaning from the instance of social organizations.

Towards making further progress on these lines, let's consider the various kinds of system to which the term 'organized' is applicable in normal usage. Apart from organisms themselves, together with their societies and institutions, the term is clearly applicable to various inanimate and conceptual structures. For example:

(1) Abstract structures: e.g. theories, musical compositions and other works of art, legal codes, grammatical sentences, etc.
(2) Skilled performances: e.g. piano playing
(3) Mechanical and electrical structures: e.g. cars, computers, beaver's dams
(4) Algorithms: e.g. computer programs, timetables, etc.

What these examples have in common, apart from being artefacts, is that they are complex systems whose component parts or entities are interconnected. Yet this statement is not yet sufficiently restrictive. As has been said, not any sort of arrangement of interrelated parts is sufficient to qualify a 'whole' as being organized. If a few random changes are made within a computer it ceases to be an organization since it ceases to be able to compute. It would seem therefore that only those artefects are acceptable as organized systems which 'work' satisfactorily – and which can thus be said to have a function!

This is the situation I have referred to where the meaning of 'organization' seems to be thrown back on the presence of functions. Yet this need not trouble us in the case of artefacts since the functions of man-made entities can be clearly specified. It is only in the case of organisms themselves that the notion of 'function' is ambiguous. Unlike an artefact an organism cannot be regarded as an assembly of previously known components put together for the purpose of achieving a specific objective such as might be laid down in the form of a patent. The study of biological systems proceeds in a manner which is the reverse of that which is adopted

** I have argued this previously (1975 b) and other authors such as Klein, Blum and Apter have made the same point although from rather different viewpoints.

by the engineer; the 'functions' of an organism are read into it on the basis of hindsight whereas the engineer creates an artefact according to his intentions in advance.

To be sure organisms are organized systems *par excellence* as it were. Therefore it would be correct, I think, to accept them as such *without* the reference to function which is needed in the case of artefacts. From this point of view the actual function (or functions) of organisms are to be regarded as *consequences* of their state of being organized; that is to say a particular living creature has those particular functions which its own special state of organization is capable of achieving for it. This would be entirely consistent with the view of many evolutionists who regard organisms as attaining new functions *subsequent to* the structural variations which make them possible.

Whereas the simplest artefacts usually have only one function, more complex organized systems may have many. In particular the higher organisms have the ability to try out various strategies in a trial- and -error manner, together with the ability to judge which of these strategies works the best in relation to the particular external circumstances. In other words they are able to make an act of selection. This, as has been seen, is what is needed for inventiveness.

What further can be said about the meaning of 'organized'? Although the term is very widely used in the biological and sociological literature there has been surprisingly little attempt to analyse it. However an interesting clue comes from an unexpected quarter – from Eddington (1935;108). He writes, in his inimitable style: "We often think that when we have completed our study of *one* we know all about *two*, because 'two' is 'one and one'. We forget that we have still to make a study of 'and'. Secondary physics is the study of 'and' – that is to say, of organization." Of course he is using the word 'organization' in a very wide sense – much wider than I have adopted above. Yet his remark is an important one; as soon as we have two or more interacting entities we have something else as well: a relationship, or a set of relationships, between them.

Ross Ashby (1962) has taken the matter a stage further and has suggested that the hard core of the notion of 'organized' is *conditionality*: "As soon as the relation between entities *A* and *B* becomes conditional on *C*'s value or state then a necessary component of organization is present". And he goes on to say that we can get a further notion of organization by asking about its converse and this converse is separability. Thus organized systems are those which exhibit the quality of non-separability and he seems to mean by this something analogous to a departure from that pairwise additivity of interactions which usually exists in mechanical systems. *

---

* An interesting example of non-separability, of the kind to which Ross Ashby may be referring, is provided by the dominance relations within a group of rhesus monkeys. It is precisely because the monkeys form *coalitions* among themselves that the dominance relations cannot be predicted from a knowledge of the interactions between pairs. (Wilson, 1975;7).

A notion somewhat similar to Ross Ashby's had been put forward previously by Woodger (1960): "The most basic principle of biology", he said, "is the statement which asserts that living things have *parts* which stand in the relation of *existential dependence* to one another." He instances the mutual dependence of a man's head and trunk; neither can survive without the other. A leg, on the other hand, is existentially dependent on the rest of the body, but not vice versa since a leg can be severed without loss of life to the remainder.

These ideas link up very readily with other aspects of organised systems which have been mentioned in the literature. For instance with the view that 'organization' implies a continuing pattern or framework, one which endures even if the 'parts' are being continually replaced, as occurs in the living body (Young, 1951;135). And again with the notions of *plastic* and *marginal* and *hierarchical* control which have been emphasised by Popper and Polanyi and Pattee respectively.

Some authors have argued that organisations necessarily have a hierarchical structure, but this would appear to be erroneous in view of the existence of certain counter-examples – e.g. colonial animals and also the brain itself. There are no present indications that any single neurone, or a small group of them, controls the whole nervous system. But of course organised systems often are hierarchical. Notice too that the notion of internal constraints (or controls) giving rise to an integrated performance is not to be taken as meaning that 'the whole', in some mysterious sense, is greater than the sum of the parts (Nagel, 1961a). It is rather a question of feedbacks, and so on, such as are very familiar in purely mechanical systems. Nevertheless the detailed knowledge of integration within the organism remains very fragmentary.

By way of a brief summary of the main points: (a) An organised system, whether living or inanimate, is the set P of all parts or components, P_i, together with the connections or relationships existing between those parts; (b) In the case of artefacts, a system could not be said to be organised unless it has a specifiable function – i.e. unless it can be said what it is organised for. This cannot be done in the case of organisms which must therefore be regarded as organised systems *de jure*, by virtue of being alive. Mere 'survival' is far too vague an idea to serve as a criterion of an organism's function (Grene, 1965). For instance it doesn't offer any means of comparing one species with another, since the various species are capable of adopting quite different methods of surviving – and different methods at different times in different niches. It is only when a particular species has significantly declined in population, and more especially when it has already become extinct, that it can truly be said to be less well organised to survive than some flourishing species. A presently flourishing bacterium appears to be just as effective at survival as is man!

§ 5. A Speculation About Life's Ongoings. The question whether or not there is a 'trend' in evolution has been much discussed by biologists and has been found to be full of ambiguity. G. G. Simpson (1950) pointed out that 'increase of complexity' is a very uncertain criterion; the evolution of some species is accompanied by structural simplification, rather than by increased structural complexity, and furthermore it is very difficult to compare the 'complexity' of one species with that of another. Asking himself whether there is any other criterion of 'progress' such that evolution could be said to be progressive, he found himself largely in agreement with Lotka who maintained that the most general law of organic evolution is that it leads to a net increase in the total mass of living matter. Simpson therefore believes that the only criterion which is sufficiently general is "the tendency for life to expand." This, he says, may be considered "in terms of the number of individual organisms, of the total bulk of living tissue, or of the gross turnover, metabolism, of substance and energy."

There is not a crumb of comfort here for the anti-materialist! But he might reasonably object that there is more to life than mere multiplication, and he could refer to the production, through man's inventive powers, of all those things which Popper (1972) has classified as World 3 entities – i.e. propositions and theories and works of art, together with the designs and principles which underly the construction of all artefacts.

It is significant that the World 3 entities are organised systems, as this term has been used in § 4. The speculation to be examined below is whether the notion of 'organised' is capable of giving expression to a one-way tendency, entirely independent of entropy increase, which would simultaneously characterise: (a) Simpson's 'tendency for life to expand'; (b) an increase of World 3 entities; and (c) an increase in the number of artefacts.

In view of the uncertainty about the precise meaning of 'organised' (an uncertainty which the previous section has done little to dispel) it may seem entirely premature to attempt any definition of the *degree* or *amount* of organisation. Yet science doesn't always proceed by moving from one well-established position to another! There can be heuristic advantages in first identifying something which is quantifiable, and then examining how well it describes whatever is felt intuitively as being in need of quantification.

Woodger (1929;291) pointed out that when cell division occurs, and when the daughter cells remain together to form a larger entity, the latter has *an increased degree* of organisation since it now has additional parts, each characterised by the spatial organisation which previously characterised the original cell. Furthermore there is a certain language usage which suggests that degree or amount of organisation might be capable of being expressed quantitatively – or at least that it should be possible to achieve a ranking order among a group of organisations of the same class. Thus we often speak of one social institution, say a business, as being 'more' or 'less' highly organized than another. And again it is sometimes said that a particular institution is 'over-organized' and this seems to imply the existence of an optimum degree of organization for a particular objective. What intuitive criteria, one wonders, does the speaker have in mind when he says these things?

Most of the published attempts at establishing a measure of organization are based on information theory and go back to a paper by Rothstein (1952). It will be recalled that Shannon's well-known measure of the information content, H, of a coded message has the form:

$$H = -\sum_{i=1}^{n} p_i \ln p_i,$$

where the p_i are certain probabilities relating to the symbols of the message. Rothstein sought to use this information measure in the context of organized systems; he accepts that the notion of organization involves the interaction of components and maintains that the greater is the interaction the greater is the degree of organization of the given set of components. He then proposes that the degree of organization of a system is the amount by which the value of H for the actual system is less than the value of H for the same components if they did not interact. In other words:

(degree of organization) = (sum of H for components taken separately
− (value of H for system allowing for interaction).

This idea was taken up subsequently by Watanabe, Hutten and others, and in particular by Lila Gatlin (1972) who applied the formula to DNA and the proteins in various species.

A vigorous criticism of the theory was made by Apter and Wolpert (1965). One of their important points is that the information theorists treat all the cells of, say, the liver as being identical and thus as being an instance of 'redundancy'. This, they say, is to overlook entirely the significance of the spatial pattern and organization of *all* the cells within the organ.

To this point of criticism let me add another: since H is additive each interaction within the system is treated as being on a par with all the others; yet to treat them thus fails to take account of Woodger's principle about existential dependence: there are *particular* interactions or connections within an organized system which are *essential* to that organization's continued existence. For example if we were concerned with the degree of organization, Φ, of a vertebrate it is necessary that the value of Φ shall fall fairly rapidly to zero if the creature's spinal chord were to be severed. Yet this is not the case with the formula proposed. Neither of the bracketed terms on the right hand side of the preceding equation would change at all sharply on the sudden death of the organism, although it is then no longer organized. The Rothstein formula thus gives little scope for recognising certain connections within an assembly, whether living or non-living, as being quite essential to that assembly having, or not having, any 'organization'.

In an alternative attempt of my own (1975 b) this notion of connectedness was made the point of departure. Consider a fertile bird's egg and ask what happens inside its shell. A chick gradually forms and this is a process in which the total 'amount of organisation' – henceforth to be called *integrality* for short – within the volume defined by the shell greatly increases. * One could hardly fail to recognise the chick as being an entity which is even more highly organised than were the original contents of the egg. Let us ask then: What else is this process of embryonic development which is easier to specify and to quantify? Well clearly it is a process in which cell differentiation occurs with the result that a great variety of organs and tissues make their appearance. And again it is a process which gives rise to a very elaborate connective network for ennabling the organs and tissues to interact with each other. Quite generally it would appear that morphogenesis is characterised, at least partially, by (a) an increasing number n of 'parts', and (b) an increasing number c of connections between those parts.

Similar considerations apply to material artefacts such as cars and computers. Although World 3 entities, such as theories and musical compositions, would be much more difficult to conceive in this way, I shall hold it out as a possibility – or at least at the merely schematic level which is my present concern.

Consider now a group of organised systems which form a particular class – e.g. the class of spring-driven watches and clocks. To specify any one of them in detail requires a specification of the individual parts, together with a specification of the entire pattern of connections between those parts. The value, Φ, of its integrality, on the other hand, does not depend on that amount of detail but requires, it is suggested, only a knowledge of the number, n, of parts, together with the number, c, of connections, and the values of certain weighting factors which take account of the relative importance, in the sense of existential dependence, of the various connections. In short it is proposed that a measure of the integrality of any member of the class is some appropriate increasing function of n and c, together with the weighting factors. If so, it should be possible to make comparisons of the Φ values within the class.

If a system has n parts, the maximum number of connections between them, allowing for 'two-way' passage, is given by $c_{max} = n\,(n-1)$. This would correspond to the completely connected graph K_n. However such a system would almost invariably be *over*-organised for its task. If every part of a computer or radio set were connected to every other part, it wouldn't succeed in doing much computing or radio wave receiving! The same point was made by Ross Ashby (1960, 1970) in relation to the degree of connectedness of neurones within the brain, and also by Simon and Lewins (1973); if a system is over-connected it becomes unstable and its regions no longer have sufficient autonomy to carry out their sub-routines effectively. Natural selection may be presumed to have resulted, in the case of living systems, in a degree of connectedness close to the optimum.

* Notice that this implies a further weakness in the information theory treatment of the meaning of 'organised'. Since the egg is a virtually closed system, its 'information content' may be expected to remain constant during the process of formation of the chick. If so, the information cannot be an adequate measure of 'amount of organisation' since this increases.

To take the foregoing beyond the schematic stage involves very great difficulties. One of the most important concerns the choice of 'base-line' at which the 'parts' are to be counted. When dealing with organisms, shall it be chosen at the level of organs and tissues, or at the lower level of cells, or at the still lower level of molecules? Although the third of these is the most fundamental, its use would result in immense practical and computational problems. And again if we are dealing with World 3 entities it is difficult to see what is the appropriate base-line for comparing the integrality of one such entity with another. As Wimsatt (1974) has rightly pointed out, complex systems can be decomposed into 'parts' in many different ways and the mappings corresponding to these different decompositions may not be isomorphic.

Nevertheless something further can be said and it concerns connectedness. There are no conservation laws of physics which require that connectedness shall remain constant. Think for instance of an organized electrical system which contains a number of switches; its connectedness increases when switches are turned on and it diminishes when they are turned off. Similarly in the case of organisms where in many cases the connections are via membranes whose permeability can be changed. Here again the connectedness is not required to be constant because of any conservation laws. Furthermore changes in the degree of connectedness will occur *at constant entropy* in a sufficiently idealized system – one in which switches, valves, etc. operate without friction or other sources of irreversibility.

In short there is no requirement that the connectedness of an isolated system shall remain constant during the course of time – it might either increase or decrease. Furthermore, if it is correct that connectedness is essential to the concept of 'organized', it seems that the degree of organization of a system, its integrality, can change *independently of entropy*. To be sure switches, valves and membranes do not operate entirely reversibly in real systems; but their associated small entropy changes would appear to be adventitious and quite unrelated to the magnitude of the change of integrality.

If my argument is not mistaken, integrality is thus seen to be a non-conserved function which is entirely distinct from that which is dealt with in the Second Law. Whilst organisms undoubtedly conform to this law, when due allowance is made for their character as open systems, they can yet display a temporal trend which is quite independent of entropy increase. Indeed at the intuitive level they display a tendency towards 'building up' which is in marked contrast to the 'levelling down' processes which prevail at the inanimate level. Of course I am not suggesting that organisms operate *contrary* to the Second Law; far from it – their irreversible processes are entropy producing like all others. But there is no reason to suppose that entropy is the *only* non-conserved function– there may well exist further functions which have not yet been adequately formulated. In my view some of the most significant processes which are studied in the life sciences (such as biological evolution, the appearance of consciousness and the creation of World 3 entities), have no direct connection with thermodynamics.

Let's turn to the question whether integrality, as well as being non-conserved, shows a monotonic temporal trend. That this is not so in the case of individual organisms seems clear. Their Φ values may be expected to rise steadily during development from the zygote, reach a plateau during maturity and then decline to zero at death (or shortly after since certain organs, such as the liver, continue to be active for a time). Material artefacts have a similar history, except that their period of decay and disutility may be much more prolonged. On the other hand it would appear that the Φ values of World 3 entities show a different behaviour. During the process of creation of, say, a mathematical theorem, a scientific theory or a musical composition, the integrality increases until the entity is complete and thereafter the integrality retains a steady value. For even though a theory may be superseded, or a piece of music becomes less popular, its structure remains intact. Its integrality thus remains constant – or at least for as long as there are human minds to understand it.

The picture is much more interesting if one considers the total integrality of *populations* of organized systems, rather than the integrality of single specimens. If Φ_i is the integrality of a specimen of the i'th class, the total integrality of the population of that class will not be less than $\Sigma\Phi_i$ where the summation is over all members of the class. (In fact it may be greater than $\Sigma\Phi_i$ due to the possibility that the members form higher level organizations, such as societies and coalitions, with each other.) If the individual members are sufficiently alike the foregoing summation may be replaced by the approximate expression $p_i\Phi_i$, where p_i is the population and Φ_i is now taken to be a mean value for the class. Suppose too that, as well as the i'th class, there are other fairly homogeneous classes whose integrality can all be reckoned relative to the same base line. Then the total integrality of the Earth's biosphere due to these classes is

$$\Phi \geqq \sum_i p_i \Phi_i$$

where the summation is now over the various classes.

Consider the time derivative:

$$\frac{d\Phi}{dt} \geqq \sum_i p_i \frac{d\Phi_i}{dt} + \sum_i \Phi_i \frac{dp_i}{dt}.$$

Clearly there is a reasonable presumption that $\frac{d\Phi}{dt}$ will be positive if a majority of the terms $\frac{d\Phi_i}{dt}$ and $\frac{dp_i}{dt}$ are positive, since the p_i and Φ_i are positive in any case.

Before asking whether these conditions are likely to be satisfied, let me revert to the inadequacies and difficulties of the present 'theory' of which I am only too well aware: (1) The notion of 'organised' is not well defined and a possible measure of it remains highly speculative; (2) It is unlikely that a common base line can be found for all classes of organised systems;

(3) The numerical evaluation of the foregoing summations would, in any case, be forbiddingly difficult. In view of the first two points especially it is not feasible to put forward a fully satisfactory argument about the likely sign of $\frac{d\Phi}{dt}$. (The third is less important in regard to a merely qualitative claim. After all the science of statistical mechanics offers many examples of qualitative statements which are made on the basis of non-computable summations.)

It seems likely nevertheless that $\frac{d\Phi}{dt}$ will indeed be positive. Within the man-made world the number of constructed entities is increasing, and thus a majority of the dp_i/dt terms due to these classes of entities may be expected to be positive. And although some of the $d\Phi_i/dt$ terms may be negative, others will be either zero or positive. In general therefore the contributions to Φ made by artefacts and World 3 systems will be in the direction of an increase.

So also in the case of the much larger contribution due to organisms. Although, as Simpson remarked, not all species have become more complex, there can be little doubt that the general level of the degree of organisation of living things has greatly increased during evolution. Recall the process in broad outline as it is believed to have occurred: first the 'biochemical soup'; then the formation of the uni-cellular organisms; the emergence of simple multi-cellular organisms during the Cambrian period; the later appearance of all the major phyla; and eventually the occurrence of creatures, such as the primates, having elaborate nervous systems. The most highly organised of all organs would appear to be the human brain which is estimated to contain some 10^{10} neurones in the cortex alone, each of them having some tens of thousands of synapses (Granit, 1977;48). All this is very familiar and I believe the reasonable conclusion is that most, although not all, of the $d\Phi_i/dt$ terms have been positive during evolution. The same can surely be said of the dp_i/dt terms. To be sure the populations of some of the simplest organisms have probably diminished since an early stage. Also a number of quite highly organised creatures, such as the dinosaurs, have disappeared. Nevertheless it is likely that a majority of the most highly organised animals and plants have increased their populations, slowly or rapidly as the case may be, since they were well endowed to meet the requirements of natural selection.

In short the result of this very rough argument is to suggest that the majority of the terms $d\Phi_i/dt$ and dp_i/dt in the foregoing inequality are positive. Therefore it seems not extravagant to conjecture that the integrality of the biosphere as a whole has increased, and is still increasing.

Of course nothing can be said about how long this trend will continue. Some great catastrophe may occur and sweep away much of the accumulated organisation. In any case the Sun's supply of nuclear fuel will probably become exhausted in time and thus give rise to a reversal of present tendencies on Earth. Life may well be a temporary phenomenon on this planet (although perhaps not in the universe as a whole). Thus I regard the conjecture of an increasing integrality as having only a very limited scope – i.e. as being applicable to the Earth's biosphere for a limited period of time.

The reader might reasonably object that the foregoing argument, as far as organisms are concerned, amounts to little more than Simpson's statement about "the tendency for life to expand". Indeed I agree. Nevertheless the picture which has been put forward brings living creatures within the scope of a more comprehensive scheme, one which includes other organised systems, such as man's artefacts and World 3 entities.

Before ending this very speculative section let me add that the model I have been using does not depend on vitalistic or teleological ideas. Leaving aside the World 3 entities (whose construction by 'mind' remains entirely mysterious), the formation of organised wholes depends on: (a) the properties of the lower level components and, in particular, their connection-forming tendencies; (b) the boundary conditions which are set by the environment. There is nothing here which cannot be understood in terms of familiar physico-chemical phenomena. Many researches on the processes of *self-assembly* have been described by Fox and Dose (1972); for example, Schmidt's experiments on the self-assembly of collagen stacks after precipitation of collagen from solution. In a significant passage, Fox and Dose remark that evolution is 'constructionist'; various components come together under the action of natural forces, and the resulting aggregate then finds for itself certain 'functions' on which natural selection can operate.

Perhaps some of my readers, whilst agreeing about the absence of vitalistic or teleological assumptions, might wish to make almost the opposite criticism: that the foregoing is altogether too mechanical, since it is based on the model of a machine and its interconnected parts. Of course this is true; but one uses an admittedly over-simplified model in the hope of making some progress. The increase of integrality in the biosphere (if I am right in thinking that it does increase) is certainly very far from providing an adequate characterisation of life's ongoings. There also occurs the emergence of sentience, and this merges imperceptibly into consciousness, and beyond that into the self-awareness and mental powers of human beings. Although science has found no means by which this wonderful process can be encompassed, my own guess, for what it is worth, is that there is yet another non-conserved function to be found here, if it can be formulated.

Chapter 8

Time and Consciousness

§ 1. Introduction. We know about 'time' from two distinct sources: from various changes and motions in the external world; and secondly and more directly, from inner experience. That there are these two sources suggests that 'time' provides a link between the physical and the mental – the theme I shall be exploring in the present chapter.

First a few words about 'mind' and 'brain'. It is in no way contrary to the naturalistic viewpoint I have been adopting to say that our own mental activity is known to us much more directly than is the physical brain. Although the mass of cells within our skulls is necessary for our thinking and feeling, the word 'brain' does not denote the thinking and feeling itself. For this the word 'mind' is used, and it is to be understood, I think, as meaning the whole sequence of our thoughts and emotions, together with whatever provides them with a unity – which makes them *our own* thoughts and emotions.

Thus, although we use the substantive 'mind', this does not refer to a thinglike entity but rather to a *process* – and, as such, time is of its essence. Thought follows thought in serial order, just as physical events do in the external world. Furthermore both sorts of events, mental and physical, define *the same* temporal order, since experienced 'presents' can be placed in a 1 : 1 correspondence with sets of external events. The concept of time thus provides a common ground.

Notice too that our awareness of time differs greatly from our awareness of space (§ 4.6). In *any one* state of awareness different things or events occupy different locations in space, but they don't occupy different times. The state of awareness is itself an event and has its temporal location within the temporal order. The 'self' is a succession of such states and is temporally extended, but is not experienced as being spatially extended.

This temporal extension of the self offers a clue to the biological utility of consciousness, a matter which has been the subject of much debate. It will be recalled that Sherrington saw the origin of consciousness in its usefulness to motor acts, whilst William James postulated that consciousness was what was needed by a nervous system "grown too complex to regulate itself".

Consider James' conjecture in relation to the most highly developed consciousness, namely in man. Does it not seem likely that one of the biological utilities of consciousness resides in allowing a person to deploy himself over a vastly extended time span? During the 'specious present', whose duration is ca. 1/10 second, sensory information is taken in from the environment. Yet humans are also able to anticipate (and sometimes accurately to predict) what is likely to happen at an appreciable time into the future. Furthermore, because of memory, humans are able to utilise vast stores of information accumulated in the past. In short, the effective extension of the present in both directions beyond the actual present offers great biological advantages and is presumably a feature of man's successful adaptation.

This aspect of his abilities is nicely illustrated by the performing of a temporally organised sequence of actions in unfamiliar circumstances – for instance the preparing of an unusual kind of meal. This requires the carrying out of a sequence of intended operations, each of them in due order and appropriately timed. Doing this is much more remarkable than carrying out a *spatial* organisation (e.g. avoiding collisions with other pedestrians) for which *sensory* clues are available. In the case of a temporal organisation the sensory clues are lacking and it requires the having of *a plan,* a higher mental faculty. The necessary sequence of actions, which may extend over quite a long period, are 'kept in mind' – or at least in broad outline, although many of the details may be subconscious.

At the level of the greater part of nature, 'time' is the dimension of change; at man's own level time also becomes the dimension of purpose, and beyond that of ethics and creativeness.

§ 2. The Unity of Consciousness. Reference has been made already to the sence of thoughts and emotions as being one's own, as belonging to oneself. What is this mysterious unity, this selfhood?

Popper (1974 a;277) has suggested that one of the factors on which the unity depends is the having of an idea of time. It needs, he says, "... an almost explicit *theory* of time ... to look upon oneself as possessing a past, a present and a future; as having a personal history; and as being aware of one's personal identity (linked to the identity of one's body) throughout this history."

To be sure, other creatures as well as man have histories. Yet man is probably alone among the organisms in *knowing* this – i.e. in knowing that he lives from birth to death. For lack of extensive memory, and other mental powers, it seems likely that animals, though having perhaps a lively sense of spatial relationships, have little sense of temporality *, but live rather as if from one moment to another. They are 'present-centred', to use a phrase of Ornstein's (1975;89). In the case of humans, on the other hand, each individual is aware of his own history and of his selfhood, and is also

* Of course this is not to deny the existence of the well-known time-keeping mechanisms in living creatures, since the mechanisms are not at the conscious level.

aware of the history and selfhood of others. This is perhaps a necessary condition for cultural evolution. Man conceives himself as having obligations towards the histories of those who will succeed him, and this implies planning for the future – even for that part of the future which will occur long after his own individual death.

The problem of the unity of consciousness can be thought of in a different way by considering man's various senses. The senses of sight, hearing and touch all give knowledge of spatial relationships, but the question arises: What unifies these relationships? C. D. Broad (1927;344) remarked that objects perceived by vision have places in a visual space, and similarly that objects perceived by hearing or by touch have places in an auditory and in a tactual space respectively; yet the three spaces, visual, auditory and tactual *have no common characteristics* and might indeed have been perceived as being three distinct sorts of space. What unifies them into a single space, according to Broad, is the inner sense of time; any one observer is able to fuse together his visual, auditory and tactual spaces because he can judge particular sights, sounds and tactual sensations as occurring contemporaneously – i.e. within his momentary 'present'. It is the awareness of the present which holds the various senses together. And so also, it may be added, it is the awareness of the present which enables a person to be confident that, when he thinks about what he is seeing or hearing, there is indeed *a correspondence* between his thinking and his sensing. He would surely fail to make any satisfactory contact with the external world if his cognition and his perception were 'out of phase' with each other – i.e. if he were not able to *unify* his mental with his sensuous experience within his 'present'.

Broad's theory thus provides a satisfying picture of the unity of consciousness at any one present. Yet the question has still to be answered why there is a *continuing* unity of selfhood; a unity over a duration, and indeed over a whole lifetime. For this one has to turn to *the function of memory,* and also to the merging of present with past during the finite duration of *the specious present.*

As Broad remarks, it is memory which enables us to bridge the gap between any two perceptions, and thereby to judge that one of them is later than the other. And of course it is also memory which enables each individual to look upon himself, in Popper's words, as possessing a past, a present and a future; as having a personal history. The disturbance to the unity of the self which occurs when there is a loss of memory is well known; it is as if that part of the self which belongs to the period before the amnesia occurred, or at least to some part of it, has vanished.

That the specious present * also assists in giving a unity to consciousness is less familiar, and is worth dealing with in some detail. The

* In his comprehensive account of the psychology of time, Whitrow (1980) remarks that the 'specious present' should preferably be called the 'mental present'. My only reservation about using this term is that other creatures as well as man appear to have a specious present. The 'psychic present' might be a suitable alternative since it is more non-committal in regard to the presence or absence of 'mind'.

following is based on a very illuminating account given by Watkins (1977). He begins by pointing out a serious error in Hume's notion of the self as being nothing more than a bundle or collection of different perceptions – perceptions which Hume regarded as distinguishable and separable, and indeed capable of independent existence. Let p_1, p_2, p_3. ... be such a bundle. "Being separate existents", says Watkins, "each of these could, logically, occur without the others. So let us suppose that p_1 occurs and then totally vanishes, then p_2, and so on, so that there is now a sequence of perfectly discrete perceptions occurring one after the other." This is not consciousness as we experience it "for the situation here would be the same, ontologically, if p_1 had occurred in one centre (say, an oyster) and then p_2 had occurred in another centre, and so on." What we understand by consciousness requires that there is a peculiar *interconnection* between successive perceptions. He quotes from Schlick to the same effect: "The individual moments of consciousness exist not for themselves but, as it were, for each other."

How does this interconnection arise? Watkins takes the example of someone proposing a toast at a dinner: "Ladies and gentlemen, let us drink a toast to Her Majesty the Queen." During this performance, which takes about ten seconds, it would be misleading to say that one has a sequence of visual perceptions running alongside a sequence of oral perceptions; on the contrary visual and oral perceptions fuse into one experience. This is similar to the point I have already mentioned as having been made by Broad, but to it Watkins perceptively adds: "As well as this integration of simultaneous perceptions from different senses, there is also integration through time. I do not first hear 'Ladies and ...', and then stop hearing that and start hearing 'gentlemen ...' The earlier perception lingers a while, merging into its successors, and fading only gradually. I am in a way still vaguely "hearing" 'Ladies and gentlemen' when he gets to 'Her Majesty ...' – it has not yet become something which, like a remark heard over the soup, now needs an act of recall for me to bring it to mind again."

The function of the specious present in 'holding together' the immediately successive perceptions thus reinforces the similar function of memory over much longer periods. And of course the two are not necessarily distinct if (as Russell suggested) the specious present is simply a particularly intense short term memory. Whether this be so or not, the specious present clearly has great biological value (c.f. § 4.6) by the very fact that it has a finite duration within which stimuli of shorter duration can be perceived in juxtaposition. In other words, the specious present provides a sufficient 'length' of message for its structure to be discerned. Moreover, since one specious present merges imperceptibly into another, the comprehension of structure is carried forward and is a rolling process, as it were. If the situation were otherwise and the specious present were of very much shorter duration, it would seem that a sequence of momentary perceptions, even though they might be recalled by the memory, could not reveal temporal structure nearly so effectively.

The foregoing picture should be capable of experimental testing, and indeed a supporting item of evidence is to be found in an article on the mind-body problem by Lord Brain (1963). He describes researches in which a subject hears an unvoiced stop consonant either as a 'p' or as a 'k'. Which of the two he hears was found to depend in part on the *succeeding* vowel. Thus although the auditory stimuli for consonant and vowel are successive, the neural state created by the first stimulus, that of the consonant, is modified by that created by the second before it enters consciousness. "Thus in the mental present", writes Brain, "there is not only overlapping, but mutual modification of the representations of events, which in physical time are successive."

It may be added that there is a certain amount of physiological evidence that there is a sort of time-keeping process in the brain, beating at the rate of about ten per second. Wiener (1955) has surmised that its function is to facilitate good organisation of the brain's work; the beat enables various memory traces to be made accessible virtually simultaneously for the purpose of guiding action at an instant.

In summary, it may be said that the unity of consciousness depends, at least in part, on the functions of memory and the specious present. These create, for each individual, a meaningful temporal structure relating to his own existence. Perhaps the reader may claim however that I am guilty of arguing in a circle; he might wish to remind me that, in Chapter 2, I *based* the concept of time on conscious experience, and now I am seeking to understand consciousness by use of temporal ideas. If this is circular, it is not viciously so. For as soon as the concept of time has been established, as in Chapter 2, it is found to be applicable to the world-at-large, and to be applicable even if man did not exist. Therefore it becomes reasonable to re-apply the time concept to a consideration of the mind's own temporal ongoing.

§ 3. The Irreversibility of Mental Processes. As has been said, the time concept has its origins in certain facts of perception and cognition. All events at the observer's own location are perceived as being within a single temporal order, an order which is capable of being intersubjectively agreed. Furthermore perception and cognition are 'one-way only' (and this goes beyond the anisotropy of thermodynamic time). Once we have seen or known something we can never *un*see or *un*know it. This is nicely illustrated in a passage from Costa de Beauregard (1963;115): "Songeons par example à l'absurdité d'une supposition telle que celle-ci: si quelqu'un connaît une théorie scientifique ou philosophique, il pourra effacer cette connaissance qu'il a de la théorie en 'anti-lisant' de la dernière à la première ligne un ouvrage imprimé à elle consacré . . ."

We are indeed observers of the world, not *un*observers, and would find it very difficult to conceive what it would be to undergo an act of *un*observing. (For of course I am not speaking in the sense of 'Venus Unobserved', but rather of a supposed reversal of 'Venus Observed'). Memory too is temporally asymmetric; we have memories of events earlier than the present, but not of events later than the present. Nor is that all; a peculiar irreversibility of memory is that it tends to proceed *forward* from some given past event, not backward. Let $\ldots e_{-n}, e_{-m}, \ldots e_{-1}$, be a set of past events I can remember. By an act of will I can fix on, say, e_{-j}, and having done so my memories proceed spontaneously forward to e_{-i}, e_{-h}, etc; they don't go backward in the sequence e_{-k}, e_{-l}, etc. To be sure, by a second act of will I can transfer to, say, e_{-l}, but then I am again presented with a 'story' in the sequence e_{-l}, e_{-k}, etc. – in other words in the sequence in which the events were actually experienced. Just why we can't easily go over our memories in reverse seems not easily explained on the basis of a purely *trace* theory; some additional mechanism must surely be active, requiring us to read off from the traces in the same temporal order as that of the events as they actually occurred.

No doubt the irreversibility of perception and cognition need not be thought of as being purely mentalistic. As suggested in § 6.6, it may well be a form of biological adaptation to the-world-as-it-is. This is by no means inconsistent with the fact that our experience is 'one-way only'.

Perhaps it may be claimed however that deductive processes provide a counter-example to the view that mental activity is always irreversible. It might be said that when we have proceeded from axioms or postulates to a conclusion, we can often reverse the argument by adopting the conclusion in place of one of the axioms and then deducing that particular axiom as a conclusion. Nevertheless this sort of formal reversibility of an argument is not an instance of a supposed reversibility of the actual mental processes. To formulate an argument in a changed sequence requires the prior existence of the original argument. The interchange of axiom and conclusion is thus necessarily later in the personal history of the particular logician, even though some other logician may have initially obtained the theorem in the reverse order.

Very interesting examples of the irreversibility of mental action are provided by problem pictures. Initially we seem to be presented by a mere collection of lines, but then quite suddenly we comprehend what the drawing represents. In more complex examples we may quickly perceive that the drawing represents, say, a duck's head facing to the left and then, after a minute or two, we find that it also represents a rabbit's head facing to the right. The important point is that once we have comprehended the drawing's significance, *we cannot withdraw* that comprehension and see the drawing once again as a mere collection of lines. The recognition of meaning is irreversible; even on looking at the drawing again after a year or more the recognition is usually immediate.

Facts such as these suggest that the processes of perception and cognition depend on the formation within the mind (or physically within the brain) of *organised connections,* and that once these connections have been formed they are seldom completely broken. A similar view was put forward by William T. Scott (1971;129) in the course of his discussion of Polanyi's theory of tacit knowledge: "... the making of a discovery, insightful or experimental, is irreversible, for once we have perceived a coherence, its clues change their character by becoming subsidiary to that coherence, and we cannot go back to the state before we perceived the coherence when the meaning of the clues was uncertain and different."

Looking at the matter from a rather different perspective, it may be said that the receiving and storage of information by the brain is indeed always an addition (i.e. an increase, in the time direction experienced as forward) and never a subtraction. (Loss of memory is not really a subtraction since it can often be restored.) Related to this is an interesting conclusion obtained by Ornstein (1969) from his experimental studies on the subjective estimation of temporal intervals. The experimental results could best be accounted for, he says, not by invoking a 'biological clock', but rather by considering the quantity of incoming information which reaches conscious awareness; the estimation of the duration of the interval is determined by the amount of storage capacity in the brain which is required to store that amount of information.

Now the notion of irreversibility applies to processes and not to objects. The irreversible character of mental activity therefore supports the view that consciousness and mind should be conceived as *processes*, and not as 'things'. Of course this conception is by no means novel. William James held it, and so have many others (e.g. van Peursen, 1966; MacKay, 1966;249, 422). It is very distinct from the notion, as held traditionally in Western cultures, in which the mind, or the soul or spirit, is regarded as thinglike; as may be seen, for example, in regard to 'survival after death' which has been taken to mean the continued existence of a discrete entity.

This traditional substance ontology, as applied to 'mind' or 'soul', was no doubt an aspect of the dualism of mind and matter, with its accompanying reification of both poles of the dualism. Rapoport (1962) remarked that the tendency to regard 'mind' as a substantive, set over against 'matter', may also have originated from the long-established requirement of European languages for every action to be attributed to an agent. 'Mind' was thus seen as the entity which is the agent of our activity.

Furthermore there had been an incomplete acceptance, up to Darwin, of the idea that man belongs within the family of all other living things. Man was held to be distinct by 'possessing' mind or soul. But of course all modern evidence goes to show that 'mind' is at one end of a spectrum of which the other end is 'sentience' – i.e. that ability to react sensitively to the environment which is present even in very primitive organisms. And again, it is known that sentience, in that meaning of the word, is present to some degree throughout the living body – or at least in those creatures which have a nervous system. Burtt (1932) quotes Henry More, the 17th century Cambridge Platonist, as having held the view that psychic activity occurs throughout the body (in sharp contrast to Descartes' ideas about the pineal gland), and this view is now supported by some of the well known results of physiological research where it has been found, for instance, that many responses are determined within the retina and optic nerve bundle, without involving the cortex. In short, the 'process view' gets rid of a concept of the mind as being an entity imprisoned within the brain, and as having no direct contact with external things. Mental activity is continuous with conscious activity and with sentience, and is not sharply localised.

Putting the point a little differently, it may be said, following William James, that in almost all contexts where the substantives 'consciousness' and 'mind' are used, it is equally satisfactory, and perhaps closer to the truth of the matter, to replace them by adjectival forms and to speak instead of 'being conscious', or as 'having conscious (or mental) activities'. This amounts to much more than a grammatical point. There is a big ontological difference between saying, on the one hand, that a person *has* consciousness and saying, on the other hand, that he *is* conscious. To be conscious, in my view, is not to be the possessor of a 'thing' which *exists,* in some sense; it is rather to be engaged in a continuing process or activity. Of course this is not to deny the usefulness of the words 'consciousness' and 'mind', any more than one could deny the usefulness of, say, 'tiredness' over and above speaking of 'being tired'. The important thing is not to mistake the substantives for physical entities.

No doubt the conceiving of mind and consciousness as processes is very far from solving the mind-body problem. It does not tell us how mental processes are related to their physical counterparts in the central nervous system. But it does succeed in putting the problem into a different, and more comprehensible, context, since it eliminates the great difficulty of Cartesian dualism concerning how two such utterly different 'things' as mind and matter could possibly interact with each other. We have well established ideas about how material things can undergo processes, and thus the mind-body problem moves into the more familiar territory concerning how a particular material thing, the brain, can display those activities we call mental. As has been said, 'time' provides the link.

§ 4. The Significance of Chance Events in Mental Activity. The following, which is a development of § 5.7, is by no means a novel conjecture, although I hope to adduce some original arguments. Put very briefly it is the hypothesis that the brain, especially the human brain, is the seat of a great profusion of chance events, and that it is this factor, together with the existence of a superb system for selection and amplification, which accounts for the brain's creative faculties.

The random character of day-dreaming, and other relaxed states of mind, provides some initial support for these ideas. In day-dreaming it is a commonplace that successive images 'pop up' in an apparently quite unrelated manner, like the successive transparencies which are displayed by a mechanical toy. Click – here is an image; click – and here is another! So also if one allows successive words to come into one's thoughts without conscious control; they appear as if thrown up in a chance fashion, perhaps depending on minute variations in the patterns of neurone firing. Of course when our attention is given to a task, successive thoughts occur (at least at the conscious level) in a much more controlled manner. Only those thoughts are attended to which serve the purposes of the task, as is shown by our ability to add on the next word or sentence *non*-randomly to what we are in process of saying or writing. Non-randomness implies the existence of rules. Naturally the *syntax* of what we are saying or writing requires the observance of rules – the rules of grammar – but the novelty and originality of its *substance* is best understood by postulating an indeterministic element, one which consists in the putting together of widely separated ideas between which there are no established rules of connection.

The association of ideas is a very old theory in psychology and has various weaknesses. In particular it conveys an over-passive impression of the character of mental activity, and thus fails to give sufficient emphasis either to emotion, on the one hand, or to rationality and control, on the other. Over and above the process of association, it needs to be supposed that mental activity, especially in the human being, includes an outstanding power of selection and amplification. That is to say, human mental activity involves the processes of picking out, and of amplifying, from what the chance processes produce, whatever may be useful, say, for a plan of action, or for the creation of a new musical composition or a new theory. This is closely related to what was said in § 3 about the sudden and irreversible realisation that there exists a coherence.

What is thus being suggested is that the mind, or brain, has a dual system of processes: (1) Indeterministic processes for the generation of thoughts or images; (2) More nearly deterministic processes for selecting and amplifying from the chance items, for evaluating them in relation to criteria, and thereby for utilising those which contribute to the solution of the self-imposed objective, the 'problem'.

The assumption that there is selection and amplification *, as well as chance, in the mental activities thus establishes an analogy with the processes resulting in the emergence of novelty during biological evolution, as discussed in Chapter 7. However there is also an important difference. For whereas biological selection depends on mechanisms, such as template action and differential reproduction, mental selection seems to depend, introspectively at least, on value judgments **. In both cases there is 'shape recognition' and the fitting of one thing to something else; a molecule to a template or to a code, an organism to an environment, and a chance idea to a value. But this need not be taken as meaning that human values are reducible to template processes, any more than it need conversely be taken (as it tends to be in Whitehead's philosophy) that the whole of nature operates according to values. Analogies are not the same as identities!

* Platt (1956) regards the presence of amplification processes within the central nervous system as having an even greater significance, for he sees them as being the origin of our sense of temporal ongoing. He points out that the notion of amplification implies temporal asymmetry – something is made bigger. "The irreversibility of our consciousness", he says, "is the same as that of our amplifiers."

** Of course these need not be fully conscious. Indeed according to Hayek (1969) the selective criteria are not so much subconscious as *super*conscious – i.e. they are above the level of conscious awareness. For a more mechanistic account of selection processes in the brain, see Laszlo (1969).

These ideas lead on to the view that the importance of the random element in the brain may have greatly increased, over evolutionary periods, through natural selection. For if indeed chance events play a useful part in providing creative opportunities, at least among the higher organisms, it may be expected that natural selection will have operated in such a way as to increase their incidence. Popper (1974 b;1058) has put forward a similar view. "I conjecture", he says, "that natural selection puts a premium on the evolution of a partial randomization of the connection (including cybernetic feedbacks) between stimulus and response." He has also made the very interesting suggestion (1978) that the origin of consciousness is bound up with the need for making choices. (See also Popper, 1972; Popper and Eccles, 1977)

Does it seem paradoxical to suggest that the most highly developed brains display chance *par excellence*? Surely not, so long as the supposed enhancement of the indeterministic element has been accompanied by an at least equal enhancement of selective power. For whereas the former, it is proposed, is essential to what, in man, is the imaginative faculty, the latter is equally essential for discrimination and reasoning.

These hypotheses are perhaps capable of being checked by experiment. For if they are true it may be expected that the most highly developed brains will display physical structures which allow of the most profuse occurrence of chance variations of neurone firing. In fact it is known that the neurones of the human brain, as well as being much more numerous than are the neurones of the brains of other primates, are much more highly interconnected by means of dendrites. So far so good, but what about sensitivity? In § 5.3 it was pointed out that the eye is sensitive to periodic patterns varying in intensity by only a few quanta. Is the brain similarly sensitive and does its sensitivity increase the higher is the evolutionary level? The answers to these questions are still largely unknown.

Early in the development of quantum theory Niels Bohr, Pascual Jordan and others conjectured that thought processes involve such minute amounts of energy that quantum limitations may play an essential role in mental activity. A good deal later MacKay (1966;466) pointed out that thermodynamic fluctuations provide a source of randomness much larger in amplitude than can be attributed to quantum effects. Therefore he thought it better to ascribe the random element, such as had already been observed experimentally in certain patterns of electrical excitation of the brain, to thermodynamic fluctuations in and around the neurones concerned.

Very thorough experimental studies which appear relevant to these issues have been carried out by Katz and his colleagues (Katz 1966). It was found that the transmission across the neuro-muscular synaptic gap depends on the 'quantal' release of acetylcholine. It should be said however that the word 'quantal', although fully justified, was used in a sense different from its usage by the quantum physicist. It refers not to energy quanta but rather to discrete 'packets' of many thousands of acetylcholine molecules contained in minute structures known as vesicles situated on the pre-synaptic surface. Responses appeared to depend on the all-or-none release of the contents of these vesicles and were thus built up additively according to whether 0, 1, 2, 3, etc., were discharged, due to the breakage of their membranes. These breakages in their turn, Katz suggested, might be due to critical collisions between vesicle and axon surfaces.

Although these results were established, for reasons of experimental convenience, at the neuro-muscular synapses, there appears to be increasing evidence that a similar mechanism prevails at the synapses within the central nervous system. Such a mechanism clearly provides great scope for the amplification of minute chance occurrences. The significant point is this: although the mechanism does not involve energy quanta or single molecules, it does involve *discrete units* of a different sort – i.e. the vesicles and their individual breakages. Therefore the logic of a single micro-event being able to 'tip the balance', between the non-occurrence or the occurrence of a macroscopic event, applies just as strongly in the present situation as it did in the examples discussed in § 5.7. Where single units are concerned, statistical law-boundedness does not apply. *

Although there is an obvious need for further experimental work, the foregoing seems to suggest that the human brain, through having such a vast number of synapses (at least 10^{14} in the cortex), may be the site of a much more profuse supply of chance events than occur in the brains of other primates. And further there may well be an optimum intensity and frequency of these events, on the lines of what was said in § 7.3. Too few chance events would correspond to a torpid and uncreative brain; too many might result in confusion, or even mental illness.

On the supposition that these conjectures will be supported by further experimental testing, let us ask whether they have any bearing on the problem of free will. As we know this problem refers to 'wantings' and 'powers'. A person may be said to do something freely only if he did it because he wanted to, and more particularly if it was within his power *not* to do it (Kenny 1975). But of course these claims do not go as far as to maintain that a person can "will what he wills". Nor do they imply that a person acts waywardly and unpredictably, for indeed we expect him to act 'in character' if only as a requirement of social life.

The free will problem is clearly at a level of description which is very different from that of science. It uses an entirely different vocabulary. Even so the scientific issue concerning determinism versus indeterminism does have a bearing on the matter. For if ontological determinism is held to be true then, in so far as a person's body is a physical entity, all of his bodily acts must be regarded as being physically determined. On the other hand, if ontological *in*determinism is held to be true, say at the level of atomic particles, this would not necessarily offer scope for free will unless a condition concerning unitary events is satisfied.

---

* I am much indebted to Sir Bernard Katz for correspondence on this matter.

This is made clear by recalling that some of the early quantum physicists suggested that the Uncertainty Principle might provide a 'chink' through which free will could be inserted, and that an important objection was raised: namely that when the effects of millions of micro-events, occurring in the brain, are added together, their total effect will be just as tightly controlled by statistical laws as it would be if each individual micro-event occurred deterministically. It will be seen however that this counter-argument falls down if the brain's operation does not always depend on the adding together of millions of micro-events. Indeed the hypothesis which has been put forward is that *single events,* whether at the level of energy quanta and thermodynamic fluctuations, or at the level of synaptic vesicles, are able to trigger off processes of selection and amplification. Thus, even though selection and amplification may occur deterministically (or quasi-deterministically, through statistical averaging), the initiating event may nevertheless be such that the *overall* process is indeterminate.

Perhaps it may be objected that if chance events in the brain are as significant as is suggested, our actions would appear much more fickle and irrational than they normally are.* Some experimental studies in psychology (Broadbent, 1973) are of relevance here. The evidence provided by these studies favoured the view that decision-making is by no means instantaneous – it extends over a certain duration. Many 'impulses' are allowed to occur subconsciously, and they do occur until a sufficient excess of impulses towards *one*, out of a number of possible decisions, has accumulated. It is only then, when a sufficient majority of 'votes' has tipped the balance, that the brain's command goes out, again without conscious awareness, to carry out that particular action. Perhaps the winning set of impulses are those which best satisfy a person's selective rules or criteria – his values.

A good indication how rules and criteria are able to produce orderly results, by filtration out of randomness, is provided by a 'model' due to Gomes (1966;451). It is an electronic music device in which single electrons, falling by chance on different sectors of a collector ring, are able to activate different amplifiers. These, in their turn, activate the various keys of a mechanical piano. However, the amplifiers do not activate the keys directly, but only through the mediation of a 'selector' which is programmed according to the rules of music, and which also has a memory for the notes which have already been struck. As a consequence a random electron impact on the collector ring does not necessarily result in a key of the piano being struck; it does so only if the corresponding musical note is consistent, according to the rules of music, with the notes which have already been played. Such a device is thus indeterminate and yet produces an orderly musical composition. Gomes suggests that the brain may operate in a similar manner and this is consistent with everything that has been said above.

---

* For example Popper (1972;233) has objected to the notion that our thoughts and actions originate in chance events in the brain on the grounds that such a mechanism could only explain snap decisions and could not explain deliberation. The considerations obtaining from Broadbent's studies show, I think, that this objection need not be decisive.

But of course the foregoing provides no evidence that we have free will; my argument is simply to the effect that science doesn't exclude it.

§ 5. Levels of Description. This brief discussion of the free will problem, very inadequate though it is, serves to illustrate an important point; that there are certain problems which require more than one level of description, or of explanation. Of course this is a familiar remark, but it has an important bearing on selection processes. The rules or criteria for selection create a linkage between different levels; for whereas the rules *operate* on a lower level, their *significance* lies at a higher one. For instance the rules which apply to the selection mechanisms within the single cell are those which best satisfy the conditions for the successful development of the whole organism. And similarly the rules which, consciously or unconsciously, affect a person's physical actions, including his acts of speech, are those which are best in accord with his wants and his values. As such they are mentalistic, rather than physiological, rules since they obtain their significance from the mental level.

This distinction needs to be enlarged on a little, as otherwise I may be misunderstood. If man's mental activity were to be regarded as a product of his body, a reductionist viewpoint, this, I suggest, would not be so much a false contention as one which is incomplete. For although mental activity depends on the brain, the outcome of this activity cannot be understood on the basis of the lower level concepts of neuro-physiology, and still less on the basis of the concepts of chemistry and physics. Towards understanding what a higher level is really about, the lower level concepts require to be *supplemented* by further concepts appropriate to the higher level – a point which has been brought out very clearly by Polanyi (1967), as well as by Grene (1971), Taylor (1971) and others.

The understanding of nature therefore requires a two-fold movement of ideas. The reductionist mode is to proceed upward – i.e. from the physically more simple to the physically more complex. We need also to make use of a downward mode, and this is particularly well exemplified by the use of notions such as 'satisfaction' and 'intention', as well as by logical concepts such as 'truth'. Towards understanding the behaviour of other animals, we may need to attribute to them (*pace* behaviourism) some of the same satisfactions and intentions as we find in ourselves. It is precisely as the type of explanation provided by physical science becomes more and more difficult, at the higher levels, that the alternative downward mode, based on man's introspective analysis of his own conscious states, becomes progressively more intelligible, and more rational.

The continuity of nature should therefore be seen as a continuity from above, as well as from below; comprehensive understanding requires the dual movement. Before I come back to 'time', let me illustrate this a little further with the example of 'truth', as referred to already. Although this example will be very familiar, from the writing of Grene and others, it has the great merit of bringing the point home within a 'hard' logical area.

The essential point is that the concept of truth, although entirely necessary to science, is not provided *from within* science. It depends not on theories of particles and fields, as is obvious, but rather on certain criteria of rationality and self-consistency. This shows reductionism to be incomplete as a hypothesis. For if a scientist were to contend that he is reducible to a system of particles and fields, however complex that system might be, he would have no grounds from within his reductionist scheme for assigning the quality of truth to that contention. This quality is applicable to propositions, but not to physical entities.

Suppose a computer were to deliver $2+3=8$. We would not enquire about the rationality of the result, for we already know it to be false. Our enquiry would not take the form of an 'argument' with the computer about the criteria of rationality and self-consistency to which we expect the instrument to work; it would rather be concerned with a search for a *defect* in the wiring or programming. On the other hand, if a friend were to obtain $2+3=8$ we would certainly wish to argue with him about the application of those criteria; our aim would be to *refute* his claim by showing it to be false.

The notions of a defect in the computer and of irrationality in our friend thus refer respectively to quite different levels of description/explanation. But of course this is not to say that our friend's body is not a physical system; it is rather to say that the concept of truth belongs to a level of discourse higher than that of physiology.

The situation is a little different with regard to the concept of time, for this provides a *bridge* between levels. As has been said, time links the mental and the physical, and is as essential in psychology as it is in natural science.

§ 6. End-piece. No doubt most scientists experience the same near-compulsion as I do – to end their books with a section on 'Conclusions' where the propositions they believe to have established are gathered together. But this, I think, needs to be firmly resisted in the case of a book on time. Very few propositions on this subject have ever been established, or are likely to be, since no vocabulary is available for the explication of 'time' apart from a vocabulary in which time is already presupposed. The words which express the most basic characteristics of time can't be eliminated, by use of Ramsey-type sentences, and be replaced by non-temporal words. In consequence 'time' cannot be explained; it just has to be accepted!

Nevertheless it may be useful to summarise the main considerations which have been put forward, and to bring certain things together.

'Time', as has been said, is not a simple or primitive concept, but is rather a construct which derives from several different sources. The primary source is the phenomena of consciousness and these include perceptual events as well as non-sensory inner experience. Indeed our understanding of the basic terms of the vocabulary of time is obtained from the temporal order of thoughts and perceptions. But of course another source is the external phenomena of change and motion, which supposedly 'give rise' to these perceptions. The time-concept must therefore be constructed in an $objective_2$ manner, so as to be applicable if man were not present in the world. 'Time' is thus neither a solely higher level concept nor a solely lower level concept, in the sense of § 5 – it is neither purely mental nor purely physical.

My theme has been the caracterisation of three somewhat different forms of the time-concept which give expression to its separable features. To be sure, the three forms are as one in regard to the most basic feature: namely, that there is a unique temporal order at each location and that 'time', in this sense, is one-dimensional. Although, throughout the book, I have tacitly adopted the relational view of time – i.e. that time is nothing more than the relationship between events – some versions of the relational theory give quite insufficient emphasis to this remarkable empirical fact about the local temporal order.

In other respects the three forms differ from each other very considerably. Time-as-experienced is associated with the sense of an ongoing, of events 'happening' and of things 'coming into being'. The moment at which we perceive events as happening we call 'the now' or 'present'. We also have the sense of events being 'earlier than', or 'simultaneous with', or 'later than' each other, and those we know as being earlier than the present we regard as 'past', whereas those supposed events which 'have not yet happened' we regard as 'future'.

From this concept, physics chooses to delete several features as being unnecessary to its purposes – notably past, present and future, together with the sense of an irrevocable ongoing. The 'time' which results when these features have been removed is the undifferentiated *t*-coordinate. Events, at any one location, are strung along this coordinate and no distinction is made between them in regard to 'presentness' or reality; they are merely said 'to occur', or 'to be', at various *t*-values. Indeed the present tenses of 'to occur' or 'to be' are the only essential temporal verbs in the abstraction presented by physics. Notice in particular that the failure of the discussion in Chaper 4 either to prove or to disprove that there is an $objective_2$ 'coming into being' of things leaves us without any $objective_2$ basis for regarding either of the two directions of the coordinate as having greater reality than the other.

The effect of including the findings of thermodynamics is to modify the theory of the *t*-coordinate and to bring the concept of 'time' a little closer to what we have from our own experience. For this science shows that the order of events, as read off along the one direction of the coordinate, is objectively distinguishable from the order as read off along the reverse direction. There is a gradient of monotonically changing entropy states, and this gradient appears to correlate completely with our own judgment of the temporal order of events, as based on our sense of 'earlier than' and 'later than'. But within the thermodynamic concept of time there is still no past, present or future – and also no 'one-way' ongoing. For although thermodynamics finds the two directions of time to be distinguishable, it does not display the one direction as being in any sense 'more real' than the reverse direction. Events still merely 'occur'. Indeed it could be deceptive even to say that things 'change' (e.g. *from* states of lower probability *to* states of higher probability), for this would be to make surreptitious use of our own sense of an ongoing. The question "Which direction along the t-coordinate is the real direction?" just doesn't arise in physical science.

To realise that there are these three forms of the time concept helps in avoiding an indiscriminate transference of attributes from one form to another, a transference which would often lead to confusion. For instance, the long-continued and arduous discussion over the foundations of the Second Law was bedevilled, in my view, by attempts to regard the matter solely in relation to the supposedly symmetrical t-coordinate, rather than in relation to the clear evidence that the universe as a whole is not t-invariant but is an evolving system – leaving aside, as has been said, the question whether the one direction of the 'evolving' is more real than the other.

Let me summarise the main differences:

Concept of 'time'	What the concept allows for: Anisotropy	A 'one-way' ongoing	Past/present/future
Theoretical physics	no	no	no
Thermodynamics	yes	no	no
Conscious awareness	yes	yes	yes

A partial reconciliation of these differences is obtainable with the help of some results of previous chapters: (a) The anisotropy, according to physics, is a statistical effect; it does not apply to atomic events, but emerges with ever greater reliability the more particles there are within the system of interest; (b) Although the human sense of a unique direction has a certain logical primacy, it may nevertheless be regarded, from the viewpoint of physical science, as conditioned by entropy-creating processes within our own bodies; (c) The corresponding naturalistic interpretation of 'now', or 'the present', is that it is a moment of heightened attention when the conscious organism prepares for 'what is about to happen'.

Perhaps the reader will ask whether one or the other of these forms of the time-concept is 'more true' than the others, or corresponds more closely to reality. In reply, it could be argued that the form of the concept which is based on conscious awareness is to be preferred, since it is the most comprehensive and does justice to the largest number of phenomena. From this point of view, the simpler forms are derived from the richest by a process of discarding; certain features of 'time', as it is to consciousness, are ignored since they are not required in the theories of physics. Against this argument the physicist might claim that the supposed process of discarding should more correctly be called one of refining. The time of physics, he might say, has been largely freed from anthropomorphic features. Furthermore it is only physical time, or rather space-time, which can provide an 'interval' having the quality of invariance, and this he regards as a necessary criterion of reality.

However my own preferred answer to the question raised above is on rather different lines. We are concerned with different levels of description, and each of the three forms serves its appropriate level satisfactorily. *This is all that is required of 'time' since it is not an existent.* It is not a 'something' whose properties are determinate and are 'out there' to be discovered. It is rather, as has been said, a purely relational concept, one which refers to the relations between events and within processes. As such, the criterion for its usefulness at any one level of description concerns its *adequacy* for the purposes of that level, and not its 'reality'.

In spite of whatever degree of clarification I may have achieved in this book, almost everything concerning 'time' still bristles with problems. For instance the question whether there is a real 'coming into being' of things or events – as the *A*-theory affirms and the *B*-theory denies – is still unanswered, and is perhaps unanswerable. Then again there is the question, which I haven't touched on at all, concerning whether 'time' should be regarded as extending to infinity in both directions, or in one direction only, or in neither.

In general it seems much more profitable to put forward conjectures and hypotheses about real temporal processes than to attempt propositions about 'time itself'. It is in the scientific and empirical area concerning *what goes on* that useful achievements can be made.

At the risk of some repetition, let me clarify what I mean by 'what goes on'. As has been said, nature doesn't provide directly any 'earlier than' or 'later than', or any 'past' or 'future'. Indeed from the standpoint of the *B*-theory, which hasn't been refuted, all events or states are equally real, and they simply coincide with particular clock readings. Yet nature does provide the gradient which has been referred to, and it is relative to a temporal direction provided by that gradient, and *conventionally chosen* to be in the same temporal direction as that of the human sense of a one-way ongoing, that one can speak of objective_2 processes or ongoings in the external world.

According to the cosmologist the most significant ongoing, in this direction of 'later than', is the steady expansion of the universe, accompanied by the burning up of hydrogen and the production of heavier elements, together with the formation of stars and galaxies, a process which will perhaps culminate in gravitational collapse. The thermodynamicist gives a corresponding significance to the process of entropy increase, and this is essentially a steady loss of the potential for further change.

On the other hand, the biologist sees the processes he is concerned with in a very different light – not as an equalisation of thermodynamic potentials, but rather as a diversification, the continued branching of the evolutionary tree with the production of ever more varied forms of life, and with a tendency for the winning out of those life forms which have the greatest reproductive power.

But one may well ask: Is this all that life is about? Many people regard modern biology with a certain feeling of disappointment, as being entirely mechanistic. But this is to neglect the remarkable significance of the state of 'being organised' which is unique to living things and their artefacts. And closely related to this, it is also to neglect the emergence of sentience and consciousness during the evolutionary process, together with the production of what Popper has called the World 3 entities. If astronomers ever succed in establishing that the universe is life-bearing to an important degree, this would greatly strengthen the view that the growth of consciousness is a significant aspect of the entire temporal ongoing – although, once again, relative to the temporal direction which we ourselves define.

It is with this conjecture in mind that I have adduced such evidence as is available in support of one of the most important features of time-as-experienced; its relationship to creativeness and purpose, and to the sense of an open future. The course of events, it has been suggested, brings genuinely new things into existence. Life itself is seen as emerging from the inanimate world by chance processes when the matter is looked at in the sequence of the lower levels upwards. On the other hand, when the matter is looked at in the reverse sense, consciousness seems to act on the world in a manner which is neither random nor deterministic, but should rather be described as purposive or intentional. We seem able to choose, although to a limited extent and only within the reachable environment, certain possibilities to become real.

Yet this is very speculative. After all that has been said, or could be said, on the subject of 'time' the matter resolves itself into a comparison of certain attitudes of mind, or intuitions. One such attitude, foreshadowed by Heraclitus, stresses the fact of change and the process of 'becoming'. During the last two centuries this tradition has developed into an evolutionary and progressive view of time. In that context it is attractive to conjecture that there occurs a steady increase of the total of conscious activity within the cosmos, a process of spiritualisation in the time direction of conscious awareness. No doubt my book has been influenced by this notion, for I see it as providing a rational ground for existence.

Yet there is a second, and very different, attitude which descends from Parmenides and Plato and Spinoza. This goes far beyond the *B*-theory, for it maintains that the most fundamental things must be timeless – or perhaps 'outside of time'. In the ultimate, it is said, there can be no earlier or later, no past or future. It was Russell, I think, who said that there is a sense in which time is an unimportant and superficial characteristic of reality. Dean Inge has similarly written: "... the overvaluing of Time is closely connected with a too exclusively moralistic view of the realm of values." In certain quietist moods one can hardly fail to find oneself drawn to this idea of a timeless world, behind or beyond our own.

But of course these attitudes or intuitions are outside the scope of this particular book. As T. S. Eliot says in his own long meditation on time:

... These are only hints and guesses,
Hints followed by guesses; ...

References

Agar WE (1943) A contribution to the theory of the living organism. Melbourne University Press, Melbourne
Alfvén H (1976) British Association meeting
Apter MJ, Wolpert L (1965) J Theor Biol 8:244
Ashby WR (1960) Design for a brain. Chapman and Hall, London
Ashby WR (1962) In: Principles of self-organization, von Foerster H, Zopf GW (eds). Pergamon
Ashby WR, Gardner MR (1970) Nature 228:784
Ayer AJ (1965) The problem of knowledge. Macmillan, London
Ayer AJ (1969) Metaphysics and common sense. Macmillan, London
Bastin T (1971) Quantum theory and beyond. Cambridge University Press
Belinfante FJ (1975) Measurements and time reversal in objective quantum theory. Pergamon
Bergmann G (1959) Meaning and existence. University Press, Wisconsin
Berovsky B (1971) Determinism. Princeton University Press
Blokhintsev DI (1973) Space and time in the microworld. Reidel, Dordrecht
Bohm D (1957) Causality and chance in modern physics. Routledge and Kegan Paul, London
Boltzmann L (1895) Nature 51:413
Boltzmann L (1964) Lectures on gas theory, Brush SG (transl.). University of California Press
Borel E (1928) Le hasard. Alcan, Paris
Born M (1949) Natural philosophy of cause and chance. Oxford University Press, Oxford
Bradley RD (1962) Br J Philos Sci 13:193
Brain Lord (1963) Brain 86:381
Broad CD (1927) Scientific thought. Kegan Paul, London
Broad CD (1938) Examination of McTaggart's philosophy. Cambridge University Press
Broadbent DE (1973) In defence of empirical psychology. Methuen, London
Brown G Spencer (1957) Probability and scientific inference. Longman, London
Buchdahl G (1969) Metaphysics and the philosophy of science. Blackwell, Oxford
Bunge M (1959) Causality. Harvard University Press
Bunge M (1967) Foundations of physics. Springer-Verlag
Bunge M (1968) Philos Sci 35:355
Burcham WE (1973) Nuclear physics. Longman, London
Burtt EA (1932) Metaphysical foundations of modern physical science, 2nd edn. Routledge and Kegan Paul, London
Capek M (1966) In: The voices of time, Fraser JT (ed). Braziller, New York
Carnap R (1966) Philosophical foundations of physics. Basic Books
Cassirer E (1956) Determinism and indeterminism in modern physics. Yale University Press, New Haven
Cassirer E (1972) In: The study of time, Fraser JT et al (eds). Springer-Verlag, Berlin
Christensen F (1976) Br J Philos Sci 28:49
Davies PCW (1974) The physics of time asymmetry. Surrey University Press
Davies PCW (1975) J Phys A8:272
de Beauregard OC (1963) Le second principe de la science du temps. Editions du Seuil, Paris
de Beauregard OC (1965) In: Logic, methodology and philosophy of science, Bar-Hillel Y (ed). North-Holland, Amsterdam

de Beauregard OC (1966) In: The voices of time, Fraser JT (ed). Braziller, New York
de Beauregard OC (1972) In: The study of time, Fraser JT et al (eds). Springer-Verlag
Denbigh KG (1953) Br J Philos Sci 4:183
Denbigh KG (1971) The principles of chemical equilibrium, 3rd edn. Cambridge
Denbigh KG (1972) In: The study of time, Fraser JT et al (eds). Springer-Verlag, Berlin
Denbigh KG (1975 a) An inventive universe. Hutchinson, London
Denbigh KG (1975 b) In: Entropy and information, Kubát L, Zeman J (eds). Academia, Prague
Denbigh KG (1978) In: The study of time, 3rd edn, Fraser JT et al (eds). Springer-Verlag, New York
Denbigh KG (1981) Chemistry in Britain 17:168
Dobzhansky T (1963) In: Philosophy of science Delaware seminar 1961-2, Vol I. Wiley-Interscience
Earman J (1974) Philos Sci 41:15
Eddington AS (1935) The nature of the physical world. Everyman edn. Dent
Ehrenfest PT (1959) The conceptual foundations of the statistical approach in mechanics (from the German edn of 1912). Cornell University Press
Eigen M (1971) Naturwissenschaften 58:465
Ellis B, Bowman P (1967) Philos Sci 34:116
Elsasser WM (1969) Am Sci 57:502
Elsasser WM (1975) The chief abstractions of biology. North-Holland, Amsterdam
d'Espagnat B (1976) Conceptual Foundations of Quantum Mechanics, 2nd edn. Reading Mass, WA Benjamin Inc
Feigl H, Meehl PE (1974) In: The philosophy of Karl Popper, Schilpp PA (ed). Open Court, La Salle Ill.
Ferré F (1970) Br J Philos Sci 21:278
Fox SW, Dose K (1972) Molecular evolution and the origin of life. Freeman
Frank P (1949) Modern science and its philosophy. Harvard University Press
Franklin RL (1968) Freewill and determinism. Routledge and Kegan Paul, London
Fraser JT (1975) Of time, passion and knowledge. Braziller, New York
Gale RM (1968) The language of time. Routledge and Kegan Paul, London
Gal-Or B (1974) Modern developments in thermodynamics. Wiley, New York
Gal-Or B (1975) In: Entropy and information, Kubàt L, Zeman J (eds). Academia, Prague
Gatlin L (1972) Information theory and the living system. Columbia University Press
Geach PT (1965) Proc Br Acad 51:321
Gibbs JW (1902) Elementary principles of statistical mechanics. Yale University Press
Gillies DA (1973) An objective theory of probability. Methuen, London
Gold T (1958) 11th international Solvay congress. Stoops, Brussels
Gold T (1967) In: The nature of time, Gold T (ed). Cornell University Press, New York
Gold T (1974) In: Modern developments in thermodynamics, Gal-Or B (ed). Wiley, New York
Gomes AO (1966) In: Brain and conscious experience, Eccles JC (ed). Springer-Verlag, New York
Goodman N (1966) The structure of appearance, 2nd edn. Bobbs-Merill
Granit R (1977) The purposive brain. MIT Press
Graves JC (1971) The conceptual foundations of contemporary relativity theory. MIT Press
Grene M (1965) Approaches to a philosophical biology. Basic Books
Grene M (1971) In: Interpretations of life and mind, Grene M (ed). Routledge and Kegan Paul, London
Grene M (1974) The understanding of nature. Reidel, Dordrecht
Grünbaum A (1969) In: Essays in honour of Carl G. Hempel. Rescher N et al (eds). Reidel, Dordrecht
Grünbaum A (1973) Philosophical problems of space and time, 2nd edn. Reidel, Dordrecht
Hall BD (1979) Nature 282:129
Hamblin CL (1971) In: Basic issues in the philosophy of time, Freeman E, Sellars W (eds). Open Court, La Salle Ill.
Hamblin CL (1972) In: The study of time, Fraser JT et al (eds). Springer-Verlag, Berlin
Hanson EA (1964) Animal diversity, 2nd ed. Prentice-Hall, New Jersey
Hanson NR (1958) Patterns of discovery. Cambridge University Press

Harré R (1965) The anticipation of nature. Hutchinson, London
Harré R (1970) The principles of scientific thinking. Macmillan, London
Harré R, Madden EH (1975) Causal powers. Rowman and Littlefield, London
Hayek FA (1969) In: Beyond reductionism, Koestler A, Smithies JR (eds). Hutchinson, London
Hempel CG (1965) Aspects of scientific explanation. Collier-Macmillan, New York
Hill EL, Grünbaum A (1957) Nature 179:1296
Hinckfuss I (1975) The existence of space and time. Clarendon Press, Oxford
Huang K (1963) Statistical mechanics. Wiley
Huntington EV (1955) The continuum. Dover
Hurewicz W, Wallman H (1948) Dimension theory. Princeton University Press
Jacob F (1974) The logic of living systems, English edn. Allen and Unwin, London
Jaynes ET (1963) In: Statistical physics, Ford KW (ed). Benjamin, New York
Joske WD (1967) Material objects. Macmillan, London
Katz B (1966) Nerve, muscle and synapse. McGraw-Hill, New York
Kelly A (1973) Strong solids. Oxford
Kenny AJP (1975) Will, freedom and power. Blackwell, Oxford
Kneale W (1949) Probability and induction. Oxford
Koestler A, Smythies JR (eds) (1969) Beyond reductionism. Hutchinson, London
Krips H (1971) Nuovo Cimento 3B:153
Lacey HM (1968) Philos Sci 35:332
Lamprecht SP (1967) The metaphysics of naturalism. Appleton-Century-Crofts, New York
Landau LD, Lifshitz EM (1969) Statistical physics. Addison-Wesley Press, Reading Pa.
Landé A (1961) In: Determinism and freedom, Hook S (ed). Collier, New York
Landsberg PT (1972) In: The study of time, Fraser JT et al (eds). Springer-Verlag, Berlin
Landsberg PT, Park D (1975) Proc R Soc A346:485
Laszlo E (1969) System, structure and experience. Gordon and Breach, New York
Latzer RW (1973) In: Space, time and geometry, Suppes P (ed). Reidel, Dordrecht
Lehman H (1965) Philos Sci 32:1
Lewis GN (1930) Science 71:569
Lewis J (1974) In: Beyond chance and necessity, Lewis J (ed). Garnstone Press
Lucas JR (1973) A treatise on time and space. Methuen, London
Luria SE (1973) Life: the unfinished experiment. Scribners
MacKay DM (1966) In: Brain and conscious experience, Eccles JC (ed). Springer-Verlag, New York
Mackie JL (1974) The cement of the universe. Oxford University Press
Margenau H (1950) The nature of physical reality. McGraw-Hill, New York
Mayer JE, Mayer MG (1940) Statistical mechanics. Wiley, New York
McTaggart JME (1927) The nature of existence. Cambridge
Mehlberg H (1961) In: Current issues in the philosophy of science, Feigl H, Maxwell G (eds). Holt, Rinehart and Winston, New York
Mehlberg H (1980) Time, causality and the quantum theory, Vol 2. Dordrecht, Reidel Publ Co
Monod J (1972) Chance and necessity, English edn. Collins, New York
Nagel E (1961 a) The structure of science. Routledge and Kegan Paul, London
Nagel E (1961 b) In: Determinism and freedom, Hook S (ed). Collier, New York
Newton-Smith WH (1980) The structure of time. Routledge and Kegan Paul, London
O'Connor DJ (1956) Br J Philos Sci 7:310
Omnes R (1971) Introduction to particle physics. Wiley-Interscience, New York
Oppenheim P, Putnam H (1958) Minnesota studies in the philosophy of science, Vol II. University of Minnesota
Ornstein RE (1969) On the experience of time. Penguin
Ornstein RE (1975) The psychology of consciousness. Jonathan Cape, London
Park D (1972) In: The study of time, Fraser JT et al (eds). Springer-Verlag, Berlin
Pegg DT (1975) Rep Prog Phys 38:1339
Peirce CS (1935) Collected works, Vol VI. Harvard University Press
Penrose O (1970) Foundations of statistical mechanics. Pergamon
Penrose O (1979) Rep Progr Phys 42: 1937
Penrose O, Percival IC (1962) Proc Phys Soc 79:605

Penrose R (1979) In: General relativity, Hawking SW, Israel W (eds). Cambridge University Press
Piaget J (1946) Le développement de la notion du temps chez l'enfant. Presses Universitaires Françaises, Paris
Piaget J (1952) L'épistémologie génétique. Presses Universitaires Françaises, Paris
Pirenne MH (1952) Nature 170:1039
Platt JR (1956) Am Sci 44:180
Polanyi M (1967) The tacit dimension. Routledge and Kegan Paul, London
Popper KR (1950) Br J Philos Sci I:117 and 173
Popper KR (1956) Nature (a) 177:538; (b) 178:382
Popper KR (1957) Nature 179:1297
Popper KR (1957) Br J Philos Sci 8:151
Popper KR (1958) Nature 181:402
Popper KR (1965) Nature 207:233
Popper KR (1967) Nature (a) 213:320; (b) 214:322
Popper KR (1972) Objective knowledge. Oxford University Press
Popper KR (1974 a) In: Studies in the philosophy of biology, Ayala FJ, Dobzhansky T (eds). Macmillan, New York
Popper KR (1974 b) In: The philosophy of Karl Popper, Schilpp PA (ed). Open Court, La Salle Ill.
Popper KR (1978) Dialectica 32:339
Popper KR, Eccles JC (1977) The self and its brain. Springer-Verlag, Berlin
Prigogine I (1978) Science 201:777
Prigogine I (1980) From being to becoming. WH Freeman and Co, San Francisco
Prior AN (1967) Past, present and future. Oxford University Press
Prior AN (1968) Papers on time and tense. Oxford University Press
Quine JVO (1976) The ways of paradox and other essays. Harvard
Rapoport A (1962) In: Theories of the mind, Scher JM (ed). Free Press, Glencoe Ill.
Reichenbach H (1956) The direction of time. University Press, California
Reichenbach H (1957) The philosophy of space and time (from the original German edn of 1928). Dover, New York
Rensch B (1971) Biophilosophy. Columbia University Press,
Rescher N (1973) Conceptual idealism. Blackwell, Oxford
Rothstein J (1952) J Appl Phys 23:281
Roxburgh IW (1977) Br J Philos Sci 28:172
Russell B (1901) Mind 1901 X:30
Russell B (1914) Our knowledge of the external world. Allen and Unwin, London
Russell B (1915) The Monist 25:212
Russell B (1919) Mysticism and logic. Longmans, Green, London
Russell B (1936) Proc Cambridge Philos Soc 32:216
Russell B (1940) An inquiry into meaning and truth. Allen and Unwin, London
Salmon WC (1975) Space, time and motion. Dickenson, Encino Cal.
Schlesinger G (1963) Method in the physical sciences. Routledge and Kegan Paul, London
Schlesinger G (1980) Aspects of time. Hackett Publ Co, Indianapolis
Schlick M (1961) Br J Philos Sci 12:177 and 281
Schrödinger E (1950) Proc R Irish Acad 53 (Sect. A):189
Schwartz F, Volk P (1977) Nuovo Cimento Lett 18:183
Scott WT (1971) In: Interpretations of life and mind, Grene M (ed). Routledge and Kegan Paul, London
Scriven M (1957) J Philos 54:727
Sellars W (1962) Time and the world order. In: Minnesota studies in the philosophy of science, Vol III, Feigl H, Maxwell G (eds), University Press, Minnesota
Simon HA, Lewins R (1973) In: Hierarchy theory, Pattee HH (ed), Braziller, New York
Simon MA (1971) The matter of life. Yale University Press
Simpson GG (1950) The meaning of evolution. Oxford University Press
Sklar L (1974) Space, time and spacetime. University of California Press
Smart JJC (1954) Analysis 14:79

Smart JJC (1963) Philosophy and scientific realism. Routledge and Kegan Paul, London
Smart JJC (1968) Between science and philosophy. Random House, New York
Smart JJC (1971) In: Basic issues in the philosophy of time, Freeman E, Sellars W (eds). Open Court, La Salle Ill.
Stannard FR (1966) Nature 211:693
Stapp HP (1979) Found Phys 9:1
Stebbins GL (1969) The basis of progressive evolution. Oxford
Suppes P (1974) In: The structure of scientific theories, Suppe F (ed). University of Illinois Press
Swinburne RG (1977) Mind 86:301
Synge JL (1956) Relativity: the special theory. North-Holland, Amsterdam
Synge JL (1960) Relativity: the general theory. North-Holland, Amsterdam
Synge JL (1970) Talking about relativity. North-Holland, Amsterdam
Taylor C (1971) In: Interpretations of life and mind, Grene M (ed). Routledge and Kegan Paul, London
Temperley HNV (1956) Changes of state. Cleaver-Hume Press
Tipler FJ (1979) Nature 280:203
Tryon EP (1973) Nature 246: 396
van Fraassen BC (1970) An introduction to the philosophy of time and space. Random House, New York
van Peursen CA (1966) Body, soul, spirit: a survey of the body-mind problem. Oxford University Press
von Weizsäcker CF (1973) In: The physicist's conception of nature. Mehra J (ed). Reidel, Dordrecht
von Wright GH (1974) Causality and determinism. Columbia University Press, New York
Waddington CH (1974) In: Beyond chance and necessity, Lewis J (ed). Garnstone Press
Walstad A (1980) Found Phys 10:743
Watanabe S (1955) Rev Mod Phys 27:179
Watanabe S (1966) In: The voices of time, Fraser JT (ed). Braziller, New York
Watanabe S (1969) Knowing and guessing. Wiley, New York
Watanabe S (1972) In: The study of time, Fraser JT et al (eds). Springer-Verlag
Watkins JWN (1974) In: The philosophy of Karl Popper, Schilpp PA (ed). Open Court, La Salle Ill.
Watkins JWN (1977) Br J Philos Sci 28:369
Wehrl A (1978) Rev Mod Phys 50:221
Weinberg S (1972) Gravitation and cosmology. Wiley, New York
Wheeler JA (1967) In: The nature of time, Gold T (ed). Cornell University Press
Wheeler JA (1968) Am Sci 56:1
Wheeler JA, Feynman RP (1945) Rev Mod Phys 17:157
Wheeler JA, Feynman RP (1949) Rev Mod Phys 21:425
Whitehead AN (1932) Science and the modern world. Cambridge University Press
Whitrow GJ (1961) The natural philosophy of time. Nelson, Edinburgh
Whitrow GJ (1980) The natural philosophy of time, 2nd edn. Clarendon Press, Oxford
Wiener N (1948) Cybernetics. Wiley, New York
Wiener N (1955) The Fawley lecture. Southampton
Williams DC (1966) Principles of empirical realism. Thomas, Springfield
Wilson EO (1975) Sociobiology. Harvard University Press
Wilson H van R (1961) In: Determinism and freedom, Hook S (ed). Collier, New York
Wimsatt WC (1972) Stud Hist Phil Sci 3:1
Wimsatt WC (1974) Boston studies in the philosophy of science, Vol XX. Reidel, Dordrecht
Winnie JA (1970) Philos Sci 37:81 and 223
Wittgenstein L (1961) Tractatus logico-philosophicus. Routledge, London
Woodger JH (1929) Biological principles. Routledge and Kegan Paul, London
Woodger JH (1960) Br J Philos Sci 11:89
Young JZ (1951) Doubt and certainty in science. Oxford University Press
Zenzen MJ (1977) Br J Philos Sci 28:313
Zwart PJ (1976) About time. North-Holland, Amsterdam

Index

The Study of Time III

Proceedings of the Third Conference of the International Society for the Study of Time, Alpbach-Austria

Editors: J.T. Fraser, N. Lawrence, D. Park

1978. 34 figures, 4 tables. VIII, 727 pages
ISBN 3-540-90311-9

This volume contains the papers which were given at the 3rd Meeting of the International Society for the Study of Time at Alpbach (Austria), which took place in August of 1976. The meeting, like its predecessors, was devoted to an interdisciplinary study of time; the contributions range from philosophical discussions, sociological, and "hard" scientific expositions. Different from the format of the two preceding volumes, this contains not only the papers, but carefully worked out comments which contribute to a deeper understanding of the problems. There are numerous illustrations, many of them of art-historical interest.

The Study of Time IV

Papers from the Fourth Conference of the Society for the Study of Time

Editors: J.T. Fraser, D. Park, N. Lawrence

1981. Approx. 28 figures
ISBN 3-540-90594-4

Contents: A Backward and a Forward Glance. – My Time is Your Time.– Issues of Beginnings and Endings. – Issues of Music and Time. – Miscellaneous Contributions. – Cumulative Index, Volumes I–IV.

Springer-Verlag
Berlin
Heidelberg
New York